T0235946

The Cambridge Nature Study Series

General Editor: HUGH RICHARDSON, M.A.

THE STUDY OF
THE WEATHER

A WINTRY SKY AND LANDSCAPE, MATLOCK.

(Snow was falling just before and just after this photograph was taken.)

THE STUDY OF
THE WEATHER

BY

E. H. CHAPMAN

M.A. (Cantab.), B.Sc. (Lond.), Fellow of the
Royal Meteorological Society

CAMBRIDGE:
AT THE UNIVERSITY PRESS
1919

CAMBRIDGE
UNIVERSITY PRESS

University Printing House, Cambridge CB2 8BS, United Kingdom

Published in the United States of America by Cambridge University Press, New York

Cambridge University Press is part of the University of Cambridge.

It furthers the University's mission by disseminating knowledge in the pursuit of education, learning and research at the highest international levels of excellence.

www.cambridge.org
Information on this title: www.cambridge.org/9781107665170

© Cambridge University Press 1919

First published 1919
First paperback edition 2014

A catalogue record for this publication is available from the British Library

ISBN 978-1-107-66517-0 Paperback

PREFACE

EVERY schoolboy has an interest in knowing whether the ground will be too soft for cricket or too hard for football, whether the rain will be good for fishing or the frost hard enough for skating. We cannot control the weather. It is little use altering the screw of the barometer or holding a hot poker near the thermometer in hopes of making things favourable for a particular game or excursion. But by understanding the weather we may hope to make the best of it and to take advantage of summer sunshine in hay-making, of thunderstorms for filling cisterns, and of winter frost for mellowing soil.

The pages that follow will show that no complete understanding of the weather can be obtained on a playground or even on a parish basis. Local worthies may be weatherwise or otherwise. Among the hills they often understand the cloud caps better than a stranger. But modern meteorology has become a world science which knows no boundaries, not even the aurora borealis on the North or the setting sun on the West. The British Empire is too small for it. Its future needs some world-wide basis of international cooperation.

Perhaps the most enduring of recent victories will prove to be the conquest of the air. This has not been achieved by reckless bravery alone, but also by skill and knowledge. We have known for years that our scholars wished to fly and that our blackboard mechanics of chalk and talk fell far short of the thrills and excitements of a trip high in air. Still the parallelogram of forces has its uses amid the stresses and strains of struts and ties. And other knowledge is needful, for just as the seaman must understand the sea so the airman must know the air.

For ages past men and women have pined for the wings of a dove. And when they have won their wings above the conquered air will they even then be satisfied with "the blue realms of nothing to sway," or will they still "raise hell"[1] because they cannot bend the heavens? If only our thoughts were as lofty as our heads are high! If only our ideals could soar on wings like the aeroplane!

> "O, dass kein Flügel mich vom Boden hebt
> Ihr nach und immer nach zu streben."

Anyhow, let us encourage the rising generation to go forward with the night behind them and the day in front.

It is in this spirit that the Editor desires to commend the work of Mr E. H. Chapman (late Captain R.E.). His manuscript and letters have come from A.H.Q., B.E.F., France, and from the Third Southern General Hospital in England. Here is a valiant attempt to do quiet work under conditions which to many people would have made it impossible to do any work at all. Let us then welcome this product

> "Of toil unsever'd from tranquillity,
> Of labour, that in lasting fruit outgrows .
> Far noisier schemes, accomplished in repose,
> Too great for haste, too high for rivalry!"

H. R.

May 1919.

---

[1] flectere si nequeo superos, Acheronta movebo, Vergil; but children please choose choicer language.

INTRODUCTION

THE present volume of the Cambridge Nature Study Series has been written chiefly to provide a series of practical exercises on weather study. The necessary explanations of the various phenomena have been made as simple as possible.

In addition to serving its primary purpose as a school-book it is hoped that the book will be acceptable as an introduction to the study of Modern Meteorology. In view of the increasing importance of Meteorology as a branch of science, no apology is needed for the appearance of a new elementary text-book on the subject.

The weather is always with us, there is always something waiting to be observed. In a climate chiefly remarkable for its unseasonable weather it would be idle to expect that the practical illustrations could exactly follow a pre-ordained course. These lessons are arranged with a certain adaptability to all times and seasons, and always bearing in mind the pressing importance of observing what you can and where you can. There is perhaps no chapter which cannot be studied and illustrated at any season of the year, and yet the more majestic phenomena of storm and cloud and sunset must be eagerly watched as rare opportunity allows. The regular routine of the lessons will involve a few minutes out of doors and a few minutes diary writing in each lesson.

Exercises and problems are continuously presented to the scholar. These are numbered consecutively and are of very varied character. In deciding how best to use them the teacher will notice that they fall into several groups distinguished by the use of initial letters. Of

Introduction

course this grouping cannot be taken as rigid. It is only intended as a clue to the best way of using each question.

- O. Outdoor observation questions requiring each scholar's own original observations.
- I. Independent individual work indoors in ink at desks.
- V. *Viva voce* questions to stimulate thought; best answered by short impromptu speeches in the class-room.
- E. Extra questions for those of exceptional energy and enterprise.
- G. Generalizations from great experience, too apt to be answered from general knowledge and hearsay instead of from genuine personal experience.
- H. Homework or evening preparation questions intended to be answered in note-books.
- L. Laboratory questions.

I am glad to have this opportunity of expressing my thanks to Sir Napier Shaw for permission to use his Syllabus of Weather Study for Elementary Schools; to Captain C. J. P. Cave, R.E., Dr W. J. S. Lockyer, Messrs H. E. Carter and J. S. Dines for permission to use photographs; to Mr G. A. Clarke for permission to use both photographs and sketches; to Messrs W. Hayes and A. G. W. Howard for preparation of drawings; and to the Council of the Royal Meteorological Society for the loan of blocks. My indebtedness to the publications of the Meteorological Office will be obvious to anyone who uses the book.

I am also glad to express my thanks to Mr C. Colledge for assistance in proof-reading, and for permission to use the photograph from which the frontispiece is reproduced.

E. H. C.

KING EDWARD VII SCHOOL,
LYTHAM.
June 1919.

CONTENTS

CHAP.   PAGE

PREFACE   v

INTRODUCTION vii

I. THE WEATHER DAY BY DAY. OBSERVATIONS OF WIND .   1

II. WHAT TO LOOK FOR IN WATCHING THE WEATHER . . 14

III. CLOUDS. THE COLOURS OF THE SKY 18

IV. FOG AND MIST. DEW AND FROST 41

V. RAIN, SNOW, AND HAIL. THUNDERSTORMS . . . 47

VI. TEMPERATURE AND HUMIDITY 59

VII. THE PRESSURE OF THE ATMOSPHERE 72

VIII. WEATHER CHARTS 86

IX. CYCLONES AND ANTICYCLONES 102

X. ANTICIPATION OF WEATHER 111

APPENDIX I. REVISION AND SUPPLEMENTARY EXER-
CISES 121

„ II. SYLLABUS OF WEATHER STUDY FOR ELE-
MENTARY SCHOOLS 124

„ III. BOOKS 127

INDEX 130

LIST OF ILLUSTRATIONS

A wintry sky and landscape . . . . *Frontispiece*

FIG.		PAGE
1.	Symbols for illustrating weather experienced . . .	3
2.	Weather, March 28th, 1916. Great gale at night . .	4
3.	Compass	6
4.	Map of British Isles showing corrections to magnetic compass for year 1918	8
5.	Plan of an observation ground	9
6.	Wind vane	11
7.	Windmill wind vane	12
8.	Cirrus cloud at high altitude	19
9.	The same cloud after it had passed overhead . . .	19
10.	Parts of long bands of cirrus	20
11.	Small cumulus of a summer day	20
12.	Typical cumulus	23
13.	Anvil-like extension of false cirrus capping cumulo-nimbus	23
14.	Stratus in waves, with cirrus above	24
15.	The same clouds two minutes later	24
16.	Lower part of nimbus	25
17.	Cirro-cumulus with small cumulus below	27
18.	Cirro-cumulus: mackerel sky	27
19.	Cirro-cumulus: speckle cloud	28
20.	Veil of cirro-stratus with strato-cumulus in front . .	29
21.	Alto-cumulus waves	30
22.	Sheet of dense alto-stratus with sun shining dimly through the cloud layer. Some dark nimbus is forming below: rain fell some hours later	31

FIG.		PAGE
23.	Strato-cumulus	33
24.	Strato-cumulus at 3000 feet, from above	34
25.	The same cloud from 1000 feet higher	34
26.	Solar halo of May 27th, 1912, as seen at Aberdeen . .	38
27.	Solar halo of March 5th, 1908, as seen at Aberdeen .	38
28.	Fog, after cold night, filling the valley through which runs the L. & S.W. Railway main line to Portsmouth	42
29.	Fog creeping over top of South Downs	43
30.	Rain-gauge	50
31.	Measurement of rainfall	52
32.	Lightning flash taken by a moving camera . . .	57
33.	Tree on Wandsworth Common, London, struck by lightning June 14th, 1914	58
34.	Stevenson screen with thermometers	63
35.	Ship's screen	64
36.	Hair hygroscope	67
37.	Wet bulb thermometer	68
38.	Diagram giving the state of the atmosphere as to dryness or dampness from readings of wet and dry bulb thermometers	69
39.	Pressure on a sphere immersed in water	74
40.	Simple barometer	78
41.	Kew pattern station barometer as mounted at South Farnborough	80
42.	Scale and vernier on barometer	84
43.	Key map of Europe showing positions of Weather Observing Stations . . . *between pages* 86, 87
44.	Map of pressure, August 28th, 1917, at 7 a.m. . . .	92
45.	„ winds „ „ „ . . .	95
46.	pressure, September 10th, 1917, at 7 a.m. . .	99
47.	„ temperature, August 28th, 1917, at 7 a.m. . .	101
48.	„ weather „ „ „ . .	104
49.	Weather chart for 7 a.m. August 28th, 1917 . . .	106

List of Illustrations

FIG.   PAGE

50. Travelling depression of August 27th, 28th, 1917

                                          between pages 110, 111

51. Weather chart for September 10th, 1917, at 7 a.m. .   .   109

52. True cirrus showing sheaf of fibres with tufted ends   .   112

53. Barograph   113

54. Veering and backing winds   115

55. Trace of a barograph at Reading, Berkshire, during pas-

        sage of depression shown in charts of Fig. 50 .   .   118

The following figures are taken from publications of the Meteorological Office: 3, 16, 20, 26, 27, 34, 37, 42, 53 from the *Observer's Handbook*; 22, 52 from *Cloud Forms*; and 35 from the *Barometer Manual*.

Figs. 2, 4, 43–51, and 55 have been prepared from information supplied by the Meteorological Office. The permission of the Controller of His Majesty's Stationery Office to make use of these diagrams and data has been obtained.

The frontispiece is a reproduction of a perfectly natural photograph taken by C. Colledge, Esq.

The photographs in Figs. 8–15, 17–19, 21, 23–25, 28, 29 are by Captain C. J. P. Cave, R.E.; the photographs in Figs. 16, 20, 32 are by Dr W. J. S. Lockyer; the photographs in Figs. 22, 52 and the sketches in Figs. 26, 27 are by G. A. Clarke, Esq.; the photographs in Fig. 33 are by H. E. Carter, Esq.; and the photograph in Fig. 41 is by J. S. Dines, Esq. The drawing in Fig. 1 is by A. G. W. Howard, Esq., that in Fig. 55 is by W. Hayes, Esq.

Blocks for Figs. 8–15, 17–19, 21, 23–25, 28, 29, 32 and 41 were lent by the Council of the Royal Meteorological Society

CHAPTER I

THE WEATHER DAY BY DAY
OBSERVATIONS OF WIND

WEATHER DIARY.

1. (V.) Describe the weather of yesterday.

You will probably find that you do not remember much about yesterday's weather. You did not take particular notice of it. Generally speaking, you do not trouble very much about the weather except on special occasions, but you feel its effects all the same.

Your first piece of work will be to commence a systematic study of the weather day by day, and to make a permanent record of what you observe.

EXERCISE.

2. (O.) In a note-book kept for the purpose write a short but complete account of to-day's weather. Continue to do this day by day, making a separate complete entry for each day.

If advisable, one weather diary may be kept by the whole class, members of the class taking a day each in rotation. The entries should be discussed as frequently as possible by the whole class.

EXAMPLE OF AN ENTRY IN A WEATHER DIARY.

Weather in North-Eastern France, September 25th, 1915 (*Battle of Loos*): Sky covered with low clouds almost the whole day. Mild. Light breezes chiefly from the South-West. By afternoon the sky became gloomy, and in the early evening drizzling rain fell. About 6 p.m. a narrow and rather faint rainbow was seen.

In addition to a properly kept weather diary you should adopt some shorter method of recording what you observe each day. One interesting way is to use a number of signs or pictures, each of which represents a certain type of day. Ten such signs are given in Fig. 1. You can use these and add others when necessary.

EXERCISES.

3. (I.) Underneath the written entry in your weather diary sketch in an appropriate weather sign for each day.

4. (I.) Prepare larger and more elaborate sketches of your weather signs, say on cards 8-inch square, or about the size of cricket score figures. Provide space on a wall of your class-room for seven of these signs. Put up a sign for each day of the week, and at the end of each week discuss the week's weather before the signs are taken down.

In connection with this exercise the signs given in Fig. 1 may be copied on a large scale.

5. (I.) Obtain copies of the entries in the weather diaries of other school-fellows for some particular day. Draw a map, mark the positions of the places from which your information has come, and sketch in the proper weather sign near to the position of each place on your map. You will thus have a chart showing the distribution of weather for that particular day (see Fig. 2).

6. (E.) In the holidays you and your school-fellows will perhaps be going to different holiday places in the British Isles about the

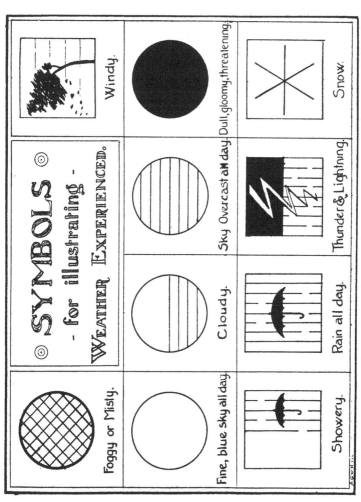

Fig. 1. For use in connection with Exercise 4.

1—2

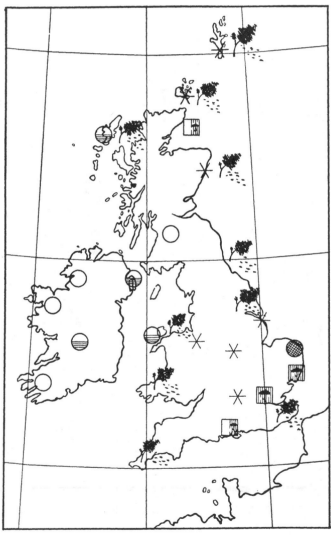

Fig. 2. Weather, March 28th, 1916 (great gale at night).

same time. If possible fix a day on which as large a number of the class as possible could take notes of the weather wherever they happened to be. Collect the observations when school reopens. Place a wall-map of the British Isles in front of the class, and fix on it your weather signs near the places on the map from which the observations have come. You will then see what the weather was over a large area on the day chosen (see Fig. 2).

WIND DIRECTION.

EXERCISES.

7. (V.) When do you feel the air, and when do you not feel the air? From your answers give a definition of wind.

8. (G.) What are the characteristics of a South-Westerly wind in our islands?

9. (V.) In a certain town the residential quarter is situated to the North-East of the factories. Why is this a bad arrangement?

10. (I.) Draw a plan of a farm-yard showing relative positions of the cow-byres and the farmer's house.

11. (H.) Write down any old sayings you have heard about wind and weather. Add to your list others obtained from your friends or from books.

> *Example.* The North wind doth blow,
> And we shall have snow.

In watching the weather great importance must be attached to the observation of the direction and strength of the wind. It is not always easy to find a suitable place at which to take wind observations. Trees, buildings, etc., may cause an observer to register wrong directions if they are near to him when observing. Wind observations should be taken in as open a position as possible. If no better place can be found the flat roof of a high building may be used.

Observations of wind direction should be taken from the sixteen points of the compass shown in Fig. 3.

EXERCISE.

12. (O.) Visit a number of open spaces in the neighbourhood of your school. Report on them as to their suitability for taking wind observations.

When you have selected your observation post you will have to determine with accuracy its orientation, that is you will have to determine the principal directions

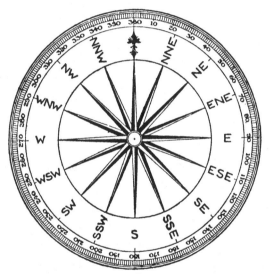

Fig. 3. Compass.

North, East, South and West. This can be done in various ways. One convenient way is to use a magnetic compass. If you use a compass you must remember that the needle does not point to true North, but, in our islands, to a point West of North. Most probably the compass you use will be divided up into degrees. There are 90 degrees in a right angle. Starting at North, and going through East,

South, West, and back to North, you will find that East is numbered 90 degrees, South 180 degrees, West 270 degrees, and North 360 degrees (see Fig. 3).

(4 right angles = 4 × 90 degrees = 360 degrees.)

Since the needle of the magnetic compass points to the West of North, when you take a bearing with the compass you will have to get the true bearing by subtracting a certain number of degrees. This number of degrees for any particular place can be obtained from Fig. 4. For London the number of degrees to be subtracted is 15, for Manchester 17, for Dublin 19, and for Valencia 20. (These values are for 1918.)

Example of use of a magnetic compass.

From a marked position in an aerodrome the bearing of a church steeple as given by a magnetic compass is 194 degrees. The correction is 15 degrees and must be subtracted.

True bearing therefore is 194 − 15 degrees, or 179 degrees, almost due South.

EXERCISE.

13. (O.) *Orientation of selected observation post by magnetic compass.*

Fix a flat stone firmly in the ground where you are going to stand to take your wind observations. Stand on this stone and, with your compass, obtain the bearings of some conspicuous objects such as church steeples, chimneys, etc. Remember to use the proper correction for each reading of your compass. From these bearings determine the principal directions North, East, South and West. Fix pegs or stones to mark these four principal directions.

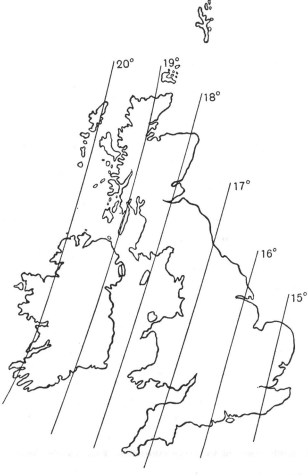

Fig. 4.   Map showing corrections to magnetic compass, year 1918.

If your wind observations are to be taken from the flat roof of a high building you could mark out the principal directions with white paint lines on the flat roof.

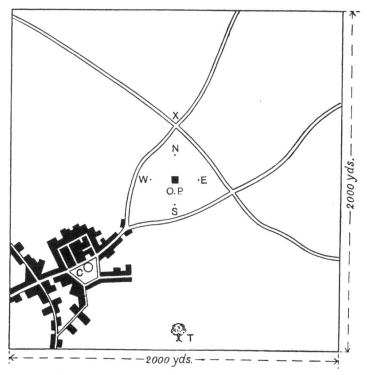

Fig. 5. Plan of an observation ground selected on a flat triangle of ground formed by three roads on the N.E. side of a town.

■   Observation Post, O.P.

N. E. S. W.   Pegs marking North, East, South, West.

C.   Church due S.W. of O.P.

X.   Cross roads due N. of O.P.

T.   Conspicuous tree very nearly South of O.P.

EXERCISE.

14. (O.) *Orientation of observation post by means of an ordnance survey map.*

Obtain an ordnance survey map of your district and mark on it the exact position of your observation post. Determine the bearings from your observation post of some prominent landmarks such as the summit of a distant hill. Draw lines N.—S., E.—W., on the map through the position of your observation post. See if you can identify any conspicuous objects along these lines. Make notes of the bearings of the objects you select.

Now that you know the principal directions on your observation ground, you will be able to obtain the direction of the wind. This is best done by facing the wind in such a position that you feel its effects equally on both sides of your face. A good method is to turn about until the wind has an equal effect on either ear. Start by placing your head so that the wind blows into one ear, and not into the other. Then turn your head round slowly until you feel the same effect from the wind on either ear. You will then be facing the wind, and, from the known orientation of your station, you will be able to give the direction from which the wind is blowing.

A good indicator of wind direction is a streamer attached to a tall pole in an open situation. Wind direction can be determined by watching the drift of smoke, or the movement of flags. The smoke, or flag, however, must not be far away from the observer otherwise errors will be made.

With a light wind some observers strike a match and note which way the smoke goes. A well-known method of obtaining wind direction with a light wind is to wet one's finger, and hold it up to the wind, and see which side feels coldest.

If you see a wind vane do not take wind directions from it unless you are quite sure that it works freely, and that its bearings are correct.

EXERCISES.

15. (O.) Obtain the direction of the wind at various times. Make a note of each observation, putting down the date, time and wind direction.

16. (O.) If possible, erect a streamer on a pole near your observation ground.

17. (O.) Observe the direction of low cloud at the times you take a wind direction. Do the two directions generally agree?

18. (L.) Make a wind vane in wood or metal, using Fig. 6 as a guide. Be very careful to construct the vane so that it moves freely on its pivot. You must also see that the vane is balanced properly. Test the vane well in the workshop by spinning it round and noting if it shows any tendency to set itself in a particular direction.

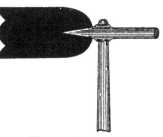

Fig. 6. Wind vane.

19. (L.) In your workshop make a windmill wind vane, using the sketch in Fig. 7 as a model. Find out where the inscription is from and read the poem.

20. (O.) Place your wind vane in a convenient position on your observation ground. Be sure that it is well exposed to the wind on all sides. If the principal directions N., E., S., and W. are indicated under your vane they should agree with the directions you have already made out in Exercises 13 and 14.

STRENGTH OF THE WIND.

In addition to observing the direction of the wind you must form an estimate of its strength. You can do this by noting the effect of the wind on various objects round about you. A numerical scale ranging from 0 calm, to 12 a hurricane, is in general use to denote the strength of the wind. This scale was devised in 1805 by a British admiral named Beaufort. Admiral Beaufort's original specifications referred to the effect of the wind on a man-of-war of his time, but equivalent specifications

Fig. 7.   Windmill wind vane.

have been devised for use on land. These land specifications are given on page 13.

EXERCISES.

21. (O.) Estimate the strength of the wind according to the Beaufort Scale whenever you obtain the direction of the wind. Make a note of both force and direction in this way: S.W., force 3, giving the day of the week, the date, and the time of observation.

22. (E.) Take a number of observations of wind direction and strength half-an-hour before and half-an-hour after sunset. Record any changes you observe.

23. (E.) When you have completed a month's observations of wind find out at what time of the day the wind was generally strongest that month.

BEAUFORT SCALE OF WIND FORCE.

Beaufort Number	Name of Wind	How to identify the force by the effect of the wind on the observer and on surrounding objects	Velocity of wind in miles per hour 5 feet above ground	Symbol for use on Wind and Weather Charts
0	Calm	Smoke rises vertically.	0	⊙
1	Light Air	Direction of wind shown by drift of smoke, but not by wind vane.	2	
2	Slight Breeze	Wind felt on face; leaves rustle; vane moved by wind.	4	
3	Gentle Breeze	Leaves and small twigs in constant motion; wind extends light flag.	6	
4	Moderate Breeze	Raises dust and loose paper; small branches are moved.	9	
5	Fresh Breeze	Small trees in leaf begin to sway; crested wavelets form on inland waters.	13	
6	Strong Breeze	Large branches in motion; umbrellas used with difficulty; whistling heard in telegraph wires.	18	
7	High Wind	Whole trees in motion; inconvenience felt when walking against wind.	24	
8	Gale	Breaks twigs off trees; generally impedes progress.	30	
9	Strong Gale	Slight structural damage occurs (chimney pots and slates removed).	36	
10	Whole Gale	Seldom experienced inland; trees uprooted; considerable structural damage occurs.	42	
11	Storm	Very rarely experienced, accompanied by wide-spread damage.	48	
12	Hurricane	Wind blows with terrific force.	over 50	

CHAPTER II

WHAT TO LOOK FOR IN WATCHING THE WEATHER

THE weather signs used in the last chapter describe, in a general sort of way, the weather experienced during a whole day. You should now go more into details when you describe the weather at any particular time. In using a weather sign you describe the most striking feature of a day's weather, and you omit the less important features. To take a weather observation properly, you must note **everything** that you observe. For example, it would not be sufficient for an observer of weather to report only " windy " at an observation hour. That would not be all he observed. He would have to report on the state of the sky, whether cloudy or not, and he would have to report on the state of the atmosphere, whether misty or foggy, or clear.

Sometimes it is possible to describe the weather at some particular time by a single word, but often it requires a combination of words.

Just as you have had signs to represent a day's weather, so you will have letters of the alphabet to describe in a conveniently short way the weather at an observation hour. The letters are easy to remember, for they are generally the initial letter of the word the letter stands for. Thus c = cloudy, r = rain, f = fog, and so on.

This way of describing weather by letters was devised by Admiral Beaufort. You have had Admiral Beaufort's scale of wind force in the last chapter.

Suppose that you are taking an observation on your observation ground. You have obtained the direction of the wind, and have noted its force. You now want to record the weather as it is at the time of observation. First of all you should look at the sky and note its appearance, using one or more Beaufort letters (see below). You may have to use two of the letters such as *og*, overcast gloomy ; or one may suffice, *c* = cloudy. Then you should look round you in all directions to see if there is any mist or haze or fog. If there were a thick fog you would be unable to see the sky, and all you could do would be to use the letter *f.*

BEAUFORT LETTERS FOR DESCRIBING WEATHER.

APPEARANCE OF THE SKY.

b = blue sky, very little cloud.

bc = patches of blue sky with detached clouds.

c = cloudy, openings between the clouds.

o = completely overcast with cloud.

g = gloomy.

u = ugly, threatening.

CLEARNESS OF THE AIR, MIST, Etc.

v = unusually clear air making distant objects unusually visible.

z = haze.

m = mist.

f = fog.

When the air is ordinarily clear no letter is used.

RAIN, SNOW, Etc.

d = light, drizzling rain.

p = passing showers.

r = rain falling steadily.

s = snow falling steadily.

h = hail.

rs = sleet.

sp = snow showers.

hp = hail showers.

rhp = showers of rain and hail mixed.

OTHER LETTERS.

t = thunder.

l = lightning.

q = squally wind.

w = unusually heavy dew.

x = hoar frost.

You should be careful to distinguish between *d* = drizzle, *p* = passing showers, and *r* = rain. The distinction is of importance when the weather over a large area is being considered.

Note that while	d = drizzle	w = dew ;
while	h = hail	z = haze ;
while	s = snow	q = squally wind ;
and while	f = fog	x = frost.

Mist is slight fog. The symbol m should be used when objects are indistinct, but traffic by road or rail is not impeded. If the obscurity is thick enough to interfere with traffic, f = fog should be used. When deciding between mist and fog imagine yourself in a motor-car travelling along a road, and judge as to whether the progress of the car would be hindered or not ; or imagine yourself to be by the side of an engine-driver on the cab of his locomotive and judge as to whether you or he could see the signals òr whether fog signals on the rails would be necessary.

EXERCISES.

24. (V.) Give the appropriate letter or letters to describe the weather at the present moment.

25. (I.) Write down any remarks you have heard to-day about the weather, and against them write, when possible, the equivalent Beaufort letters.

26. (V.) An observer walking across a grass lawn one morning noticed that his boots were very wet. The sky was perfectly clear at the time: what two letters would he use to describe the weather if he knew there had been nò rain in the night?

27. (I.) Give a description in your own words of the following sequence of events:

Time	Weather	Time	Weather	Time	Weather
Noon	b	1.30 p.m.	o	2.38 p.m.	tl
1 p.m.	bc	1.50 p.m.	og	2.40–3.20 p.m.	tl, heavy r
1.15 p.m.	bc	2.25 p.m.	ou	4 p.m.	bc

28. (O.) Describe "present weather" whenever you take a wind observation. Enter the Beaufort letters in your note-book.

29. (I.) Get as many of your school-fellows as possible who come in to school from different places in your district to observe the weather at 7 o'clock one morning. Collect the observations in morning school and draw a local chart of weather for that morning.

The Beaufort letters can be used to describe the weather during the interval between two successive observations. When you take observations in future, write down as full a description as possible of past weather, that is the weather since your last observation. You should now enter your observations of wind and weather in systematic form as in the example given below. Do not prepare many pages of your observation book in advance, because additions will be made to your observations from time to time.

EXAMPLE. *Weather in London, April 14, 1916.*

Date	Day of week	Hour of observation	Wind		Weather		Remarks
			Direction	Force	At time	Since last observation	
April 14th	Friday	7 a.m.	WNW	4	*b*	*o, c, bc*	
1916		1 p.m.	W	3	*c*	*bc, c, bc, cprhs*	Heavy sleet and hail shower, 12.45 p.m.
		6 p.m.	WNW	4	*bv*	*c, bc, b*	*t* 1.7 p.m.

EXERCISE.

30. (O.) Take observations of wind and weather daily at 7 a.m., 1 p.m., and 6 p.m. Enter the results in your note-book in the manner shown in above example.

CHAPTER III

CLOUDS. THE COLOURS OF THE SKY

EXERCISES.

31. (G.V.) Describe the clouds which you remember to be associated with (i) bright, sunny weather, (ii) wet weather, (iii) thunderstorms.

32. (G.V.) Write down any popular names you know for clouds.

Study the clouds attentively whenever you are able. You will soon notice that certain types are constantly recurring, and you will want to give them names, or learn the names that have been given to them. You need hardly be told that clouds are associated with weather, and that sometimes clouds give indications of coming weather. When you are studying clouds always do so in conjunction with weather both present and future.

EXERCISE.

33. (O.) When you take observations of wind and weather, and on as many other occasions as possible, note the forms assumed by the clouds. Give the colour, structure, and height. Make a note as to the speed with which they are moving. Use such words as these in your descriptions:

Colour. White, grey, bluish-grey, dark, black.

Structure. Threadlike, feathery, small flakes, large flakes, rolls, waves, ripples, clear-cut edges, ragged edges, ill-defined edges, sheet covering a small part, a large part, or the whole of the sky.

Height. Very high, high, low, very low.

Movement. Stationary, moving slowly, moving quickly.

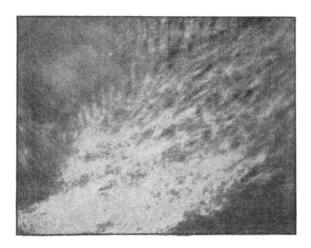

Fig. 8. Cirrus at high altitude.

Fig. 9. The same cloud after it had passed overhead.

Clouds

Fig. 10.   Parts of long bands of cirrus.

Fig. 11.   Small cumulus of a summer day.

Examples of description of Clouds.

(i) *White feathery clouds, very high, almost stationary.*

(ii) *Dark clouds with no particular shape, ragged edges, very low, moving quickly. Rain falling steadily from the clouds.*

(iii) *"...the precipices of the chain of tall, white mountains that girded the horizon at noon yesterday."*

(iv) *" Who saw the dance of the dead clouds when the sunlight left them last night, and the west wind blew them before it like withered leaves ? "*

(v) *" Those colossal pyramids, huge and firm, with outlines as of rocks, and strength to bear the beating of the high sun full on their fiery flanks...."*

(vi) *"...those war clouds that gather on the horizon dragon-crested, tongued with fire...."*

(iii) to (vi) are from Ruskin, *Frondes Agrestes.*

After you have taken a number of cloud observations you will find on looking through your notes that certain kinds of clouds occur frequently. You will pick out and put by themselves those very high, feathery, pure white, fine weather clouds. You will want to give them a special name, or learn the name which has been already given to them. An experienced cloud observer would call them **cirrus** clouds. Cirrus is a Latin word meaning a curled hair.

You would know that those lumpy, heaped-up clouds with sharp, well-defined edges described in examples (iii) and (v) above had a special name. They are called **cumulus** clouds, the Latin word cumulus meaning a heap.

For clouds that are formed in horizontal layers or sheets, you can use the word **stratus**, which means sheet or layer.

There is another cloud with which you are probably acquainted. That is the rain-cloud, called **nimbus**. The name nimbus is given to a low cloud with ill-defined edges from which steady rain is falling.

You have now four cloud names, **cirrus**, **cumulus**, **stratus** and **nimbus**. These four names were used in the classification of clouds first given in 1803 by Luke Howard of Ackworth, and the same four names are to be found in all modern cloud classifications.

Cirrus clouds, the highest of all the clouds, are composed of ice particles. The lower clouds are formed of water-drops. Cirrus clouds are very varied in shape. Sometimes they have a brushed-out, wispy appearance. They are then popularly known as "mares' tails." Often they look like long feathers, and at times a band of cirrus may be seen stretching right across the sky.

Cumulus clouds are the most common and best known of cloud forms. The way in which they are formed can be illustrated in a simple manner.

EXERCISES.

34. (L.) Boil some water in a kettle. Observe very carefully the steam coming out of the spout. Can you see the steam right close up to the mouth of the spout?

35. (O.) When you next see a locomotive on a railway note how far away from the funnel you can see the steam. If possible, watch a locomotive on a few dry, fine days, and also on a few damp, dull days. Try to find out if there is any difference in the distance you can first see the steam away from the funnel on these two types of days. If you have a camera try to obtain photographs.

In these two exercises you have something of the same sort of thing which occurs in the formation of a cumulus cloud. Strictly speaking, steam itself is invisible.

Fig. 12.   Typical cumulus.

Fig. 13.   Anvil-like extension of false cirrus capping cumulo-nimbus.

Fig. 14. Stratus in waves, with cirrus above.

Fig. 15. The same clouds two minutes later.

What you really see in these exercises is a cloud of water-drops. The steam which issues from the kettle spout is invisible. A short distance away from the spout the steam begins to condense into minute drops of water. It is a cloud of these tiny water-drops which you can see.

Cumulus clouds are formed at the top of a rising current of moist air. The upward current of air contains invisible water-vapour, like the invisible steam from the

Fig. 16. Lower part of nimbus.

boiling kettle. As the air rises it cools, and it reaches a point where the water-vapour begins to condense, and there you have a cloud beginning to form, just as the cloud begins to form some distance away from the kettle spout.

EXERCISES.

36. (E.) When a favourable opportunity occurs note the time of day at which cumulus clouds begin to form in a clear sky. Note at what time of the day the amount of cumulus cloud is greatest. Try

to obtain the time at which the last trace of cumulus disappears from the sky. If you are able to carry out this exercise completely a number of times try to deduce indications of coming weather given by cumulus clouds.

37. (E.) If the amount of cumulus increases as the day goes on, particularly towards sunset, and if the clouds at sunset darken and sink downwards what sort of weather are you likely to have the next day?

38. (E.) When you have sufficient observations of stratus clouds give the usual colour of very high stratus, high stratus, and low stratus.

39. (I.) What type of cloud is that described in Example (ii), page 21?

40. (I.) Explain the proverb "Every cloud has a silver lining." Which type of cloud often shows a "silver" edge?

The four names, cirrus, cumulus, stratus and nimbus, will not be sufficient to describe all the clouds you observe. It is somewhat difficult to provide names for all the clouds. Perhaps the best thing for you to do will be to stick as closely as possible to the classification of clouds adopted by the Meteorological Office, London, for observers in these islands.

You have four cloud names. When you are quite sure of the clouds to which they refer, make a note of any cloud you see which cannot be named by either of the four cloud names you are using.

You will sooner or later observe a very beautiful type of cloud consisting of small speckles or flocks of pure white cloud. Sometimes this cloud is arranged in ripples or waves. It is then known as **mackerel sky**. To this cloud has been given the name **cirro-cumulus**. It is a high cloud, and it is perhaps appropriate to use the word cirrus in naming it since all cirrus clouds are high, but it has no connection with cumulus except in name.

Fig. 17.   Cirro-cumulus with small cumulus below.

Fig. 18.   Cirro-cumulus : mackerel sky.

Another high cloud is **cirro-stratus**. This cloud is
not so easy to identify as is cirro-cumulus. Sometimes
you would not know cirro-stratus was there were it not
for the fact that the very thin sheet of cloud gives a
slightly milky appearance to the blue of the sky. At
times cirro-stratus is thick enough to make the whole
sky white, and give the sun or moon a watery look.

The height of cirrus, cirro-cumulus, and cirro-stratus
clouds is generally stated to be about 5 or 6 miles.

Fig. 19.   Cirro-cumulus: speckle cloud.

EXERCISES.

41. (O.) On a bright afternoon when high cirrus clouds are visible
watch the cirrus carefully, and note any changes that take place.
Repeat this exercise a number of times, and state into what other
type of cloud cirrus sometimes changes.

42. (V.) Suppose you were looking down on a big flock of sheep
from an aeroplane what type of cloud would it remind you of?

NOTE. If you find any difficulty in proceeding with your study of the clouds, or if the weather happens to be bad and you are unable to observe the clouds frequently, read through the remainder of this chapter, go on with the other work in the book, and turn back to this chapter whenever you have an opportunity of continuing your cloud studies.

Fig. 20.   Veil of Cirro-stratus, with Strato-cumulus in front.

You have now had the highest clouds, cirrus, cirro-cumulus, and cirro-stratus, described. The next highest clouds are known as the **alto** clouds. There are two, **alto-cumulus**, and **alto-stratus**. Alto-cumulus is something like cirro-cumulus in form, and alto-stratus is very much like cirro-stratus. The height of the alto clouds is from 2 to 5 miles.

Alto-cumulus, like cirro-cumulus, is a very beautiful cloud. It consists of fleecy groups of cloudlets, somewhat coarser in structure than cirro-cumulus. The

Fig. 21.   Alto-cumulus waves.

separate cloudlets are dark in the middle. The little cloudlets of cirro-cumulus are not dark in the middle. They are pure white all over. Alto-cumulus is sometimes made up of groups of separate cloudlets, and sometimes

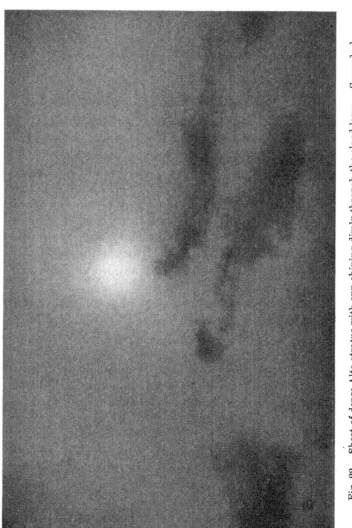

Fig. 22.   Sheet of dense Alto-stratus with sun shining dimly through the cloud layer.   Some dark nimbus is forming below : rain fell some hours later.

it is formed in waves covering the greater part of the sky. It is a very beautiful cloud to see at night when there is a bright moon. The French call cirro-cumulus *moutons* (sheep), and alto-cumulus *gros moutons* (big sheep).

Alto-stratus is a thick sheet of a dark grey colour which may cover the whole or part of the sky. The sun or moon when visible through this cloud looks " watery."

The next cloud you should learn to identify is **strato-cumulus.** It is perhaps the most common cloud of moderate height. Sometimes strato-cumulus resembles alto-cumulus, but it is darker in colour and bigger in appearance. Rolls or waves of strato-cumulus often cover the whole sky. At times small patches of blue sky may be visible between the rolls or waves. Strato-cumulus is seen most during quiet weather in the winter. It may then persist for days at a time, the weather being dull and rainless.

The following paragraph is taken by permission from a paper entitled *The Form of Clouds,* by Captain C. J. P. Cave, R.E., in the *Quarterly Journal of the Royal Meteorological Society,* January 1917, page 69 :

"Strato-cumulus is curiously thin although so persistent; it is often only two or three hundred feet thick. It is one of the most impressive experiences to pass through these clouds from a grey day on the earth's surface to the brilliance of the region above them. They appear like a wet fog when one is in them, and as one reaches their upper layers the sun struggles through, as on the earth's surface when a foggy morning is about to give place to a fine day. Suddenly the aeroplane emerges on the upper side of the cloud and a truly marvellous prospect is unfolded; as far as the eye can see great billows of cloud extend (see Fig. 24); here and there hummocks rise out of the general level, but as a whole the surface extends away in a great billowy horizontal plane (see Fig. 25). The clouds look curiously solid however, and remind one of pictures of landscapes in the polar regions. As a pilot remarked, 'They look as though one could get out and walk about on the top of them.'"

Fig. 23. Strato-cumulus.

You have used the word **stratus** to define a sheet or layer of cloud. You must now restrict the use of the word to uniform layers of cloud resembling fog, but not resting on the ground. Stratus often forms under other clouds when rainy weather is approaching.

You should continue to use the word **nimbus** to denote a cloud from which rain or snow is falling steadily.

Fig. 24. Strato-cumulus at 3000 feet, from above.

Fig. 25. The same cloud from 1000 feet higher.

EXERCISES.

43. (I.) Obtain some lantern slides, photographs, or pictures in which clouds are shown. Identify the cloud forms.

44. (O.) If possible visit an Art Gallery and discuss cloud forms as depicted in any of the paintings, *e.g.* Turner's in National Gallery.

45. (I.) Mix ink and water to make a grey colour wash. Then sketch stratus clouds on white blotting paper.

46. (E.) Continue previous sketch using red ink for sunset effects.

47. (I.) Use white blackboard chalk and brown paper. Draw high cirrus clouds.

THE COLOURS OF THE SKY.

EXERCISES.

48. (V.) Make up a list of occasions on which there are beautiful colours in the sky.

49. (V.) When does the sun look like a fiery red ball?

50. (O.) When you are taking observations, and also at other times, note any colour phenomena which occur in the sky.

The colour of the sky itself, varying from deep blue to whitish-blue is a thing of interest which may possibly give indications of coming weather. The most beautiful sky colours are seen at sunrise and sunset. These colours also serve as indications of coming weather, for example :

> Evening red and morning grey,
> Help the traveller on his way.
> Evening grey and morning red,
> Pour down rain upon his head.

EXERCISES.

51. (O.) Observe the colours of the sky at sunset on a cloudless evening. Describe as fully as you can what colours you see.

52. (O.) Observe the coloration of clouds at sunset and sunrise. Note whether different cloud forms assume different colourings.

THE RAINBOW.

EXERCISE.

53. (G.) When do you see a rainbow? With what kind of a background does it show up most clearly? What type of cloud is always present when a rainbow is formed? In what position is the sun with respect to the rainbow?

You should always pay particular attention to any rainbow you may see. You should note what colours are distinct, and which colour is the most distinct. The colours of a rainbow are sometimes said to be red, orange, yellow, green, blue, indigo, violet. But you must find out for yourself whether this is true to fact.

Sometimes only one bow is visible, but at other times two bows are visible. When only one bow can be seen the red arch is the uppermost, then come the orange, yellow, green, blue, indigo and violet arches, the violet arch being the innermost. If a second bow is seen the colours in it are reversed, the violet arch being uppermost, and the red innermost.

EXERCISES.

54. (O.) When you next see a rainbow or rainbows, note by careful observation (i) the order of the colours commencing with the innermost, (ii) the most distinct colour or colours, and (iii) the widest colour band.

55. (O.) Try to see a rainbow when the sun is shining on the spray from a garden hose, waterfall, or fountain. You will have to stand some distance away from the spray, and you must have your back to the sun.

HALOS AND CORONAS.

The sun in the daytime, and the moon at night are occasionally surrounded by rings of light. Sometimes these rings of light are coloured as beautifully as the rainbow. A small ring close to the sun or moon is called a corona. If the ring is round the sun it is called a **solar corona**, if round the moon it is called a **lunar corona**. The colours of a corona are not always distinguishable, but if you could see them you would find that the colours of the rainbow were there, the red being outermost, and the innermost colour band, violet, touching the sun or moon. Solar coronas are not so easily seen as lunar coronas on account of the strong light of the sun.

EXERCISES.

56. (O.) Obtain a piece of glass, moisten it by breathing on it, look through it at a strong light. You will see a coloured ring like a corona.

57. (O.) Try to obtain the same effect as in the last exercise by looking at a street lamp through the "steamed" windows of a tram-car on a misty night.

A halo is a ring of light which appears some distance away from the sun or moon. A lunar halo has so little colouring that it appears white. A solar halo shows the colours better, and, if you can distinguish them, you will find that they are placed in exactly the opposite way to the colours in a corona. In a halo it is the red which is innermost, and the violet which is outermost. Halos are only produced when the sun's rays, or the moon's rays, pass through a thin cloud of ice crystals. They are sometimes taken to be a sign of coming bad weather

EXERCISE.

58. (O.) Whenever you see a halo or a corona make careful notes of the time, the state of the sky, the kind of cloud present, the colours you can distinguish, and the order of the colours commencing with the innermost. Draw a sketch of the phenomenon.

Fig. 26.   Solar halo of May 27, 1912.   Complete circular halo, with
arc of contact.   (From a sketch.)

Fig. 27.   Solar halo of March 5, 1908, with arc of contact, mock sun
ring, and mock suns (parhelia).   (From a sketch.)

You should provide yourself with a piece of smoked glass, or a pair of spectacles of smoked glass for observations of halos and coronas. The strong light of the sun will not then prevent your seeing the colourings. These glasses may be used for observations of clouds in the daytime.

In connection with halos there are other colour phenomena such as "arcs of contact" and "mock suns." You will find these described in the *Observer's Handbook of the Meteorological Office, London.*

There are other colour phenomena for particulars of which you will have to refer to the more advanced books on Meteorology. There is the aurora which is seen occasionally in our islands. There is the horizontal rainbow which is formed by the sun's rays falling on drops of dew on the grass.

EXERCISES.

59. (H.) Describe any faint light you see in the sky at night, colour, shape, brilliance, position, etc.

Do you consider the light just described as due to any of the following causes? If so, which? (*a*) Milky Way, (*b*) lingering twilight, (*c*) coming moonrise, (*d*) lunar halo, (*e*) lunar rainbow, (*f*) sheet lightning, (*g*) zodiacal light, (*h*) aurora borealis, (*i*) searchlights, (*j*) distant city, (*k*) lunar corona. Give reasons for your opinion.

60. (V.) Why did Shelley describe the rainbow as so many coloured:

"The triumphal arch through which I march,
With hurricane, fire, and snow,
When the Powers of the air are chained to my chair,
Is the million-coloured bow."

61. (O.) Look at the rainbow again when one is visible. Be quite sure that you are not prejudiced by having heard other people's description of it. Say how many different colours you yourself actually

see. Are they distinct or do they shade off into each other? Are any more marked than others? Which colour is inside the bow, which outside? How often is each colour repeated?

You are now in a position to look through the following cloud classification, and to refer to it when you are taking cloud observations.

CLOUD CLASSIFICATION.

Upper cloud layer, 5—6 miles.
Clouds composed of ice crystals.
Halos seen.

Cirrus : Mares' tails ; wisps or lines of pure white clouds.

Cirro-cumulus : Small speckles and flocks of white cloud; fine ripple clouds ; mackerel sky.

Cirro-stratus : A thin sheet of tangled web structure sometimes covering the whole sky. Watery sun or moon.

Middle cloud layer, 2—5 miles.
Clouds composed of small drops of water.
Coloured rings (coronas) seen quite close to the sun or moon but never halos.

Alto-cumulus : Somewhat similar to cirro-cumulus, but the cloud masses are larger and darker in the centre.

Alto-stratus : Very like cirro-stratus, but a thicker and darker cloud.

Lower cloud layer, below 2 miles.

Cumulus : Woolpack clouds. Clouds with flat bases and with considerable vertical height. Cauliflower-shaped tops.

Strato-cumulus : Rolls or waves covering the whole or part of the sky.

Stratus : A uniform layer of cloud resembling a fog but not resting on the ground.

Nimbus : Shapeless cloud from which falls continuous rain or snow.

Two other cloud names which are in general use may be given here.

Fracto-cumulus: Small cumulus with ragged tops.

Cumulo-nimbus: (Anvil-, Thunder-, or Shower-Cloud). Towering cumulus with the top brushed out in soft wisps or larger masses (known as false cirrus), and rain-cloud at base. (See Fig. 13.)

CHAPTER IV

FOG AND MIST. DEW AND FROST

IN Exercises 34 and 35 you had examples of two very important changes which are continually taking place in the atmosphere. When water is boiled inside a kettle or inside the boiler of a locomotive the water is changed into water-vapour. Such a change is called **evaporation**. Outside the kettle the invisible water-vapour is changed into water-drops, the multitude of minute water-drops forming a visible cloud. This change, which is the exact opposite of evaporation, is called **condensation**.

Water-vapour in its invisible form is being supplied continually to the air around you. Evaporation takes place on the surfaces of rivers, lakes and seas. The surface of the ground is often moist, and evaporation takes place there, as well as from any pools of water that may have collected. Thus there is always plenty of water-vapour in an invisible state in the air. The various ways in which this water-vapour condenses and becomes visible give you some extremely important things to consider in connection with your study of the weather.

62. (V.) Why is it that you see your breath sometimes and not at others ?

63. (V.) What important form of condensation of water-vapour in the air have you already dealt with ?

64. (E.) Describe the way in which a cumulus cloud is formed.

Fog (or mist) is one of the forms of condensation of water-vapour. Whenever a fog or mist occurs the air must have contained a lot of water-vapour, and much of

Fig. 28.   Fog, after cold night, filling the valley through which runs the L, & S.W. Ry. main line to Portsmouth.

this water-vapour must have condensed into minute water-drops. A fog or mist is exactly like a cloud, in fact you could describe either as a surface cloud.

EXERCISES.

65. (V.) Suppose you were to divide all the things you have observed in your study of the weather into two groups, (i) pleasant or

agreeable, (ii) unpleasant or disagreeable, in which group would you put fog, and why ?

66. (V.) Where are fogs most frequently formed ?

Fogs are often formed by the mixing of two slowly moving currents of air, one of which is warmer and damper than the other. The warm air is able to keep its water-vapour in the invisible state until it is cooled by mixing with the colder air. Then some of its water-vapour condenses and a fog is formed.

Fig. 29. Fog creeping over top of South Downs.

A fog forming slowly in a river valley is a sight of some beauty. The air near the river has as much moisture in it as it can carry. Colder air slips down the sides of the valley, cools the damp air over the river and so causes its moisture to condense and form a fog. It is often possible to watch this fog form slowly, following the many windings of the river quite faithfully, and in time filling up all the lower levels of the valley.

A fog may be formed when a warm, damp current of air passes over a cold surface. An excellent example of this is the cloud known as the "table-cloth" which forms on the top of Table Mountain at Cape Town, South Africa. A warm, damp current of air in passing over the cold mountain-top is cooled, and some of its moisture is condensed into a fog. When the air passes away from the top of the mountain it mixes with warmer air, and its moisture becomes invisible once more.

The same kind of thing has been seen on the top of Ben Nevis in Scotland. Even when a strong wind was blowing the fog formed on the top of the mountain and remained there.

The difference between mist and fog is not great. You know which word to use in describing weather by the Beaufort letters of Chapter II. In a mist the water-drops are larger than in a fog. Mists are more often experienced in country places than in large cities.

You have now considered two forms of condensation, clouds and fog (or mist). In both of these the minute water-drops float about in the air. Cloud particles are so small that their diameters are less than one-thousandth of an inch. You may wonder how it is that these water-drops do not fall to the earth, but you have often seen particles of dust, heavier than cloud particles, floating about in the air when a bright beam of sunlight shines into a darkened room.

Fog particles are also about one-thousandth of an inch in diameter. Mist particles are larger and they feel wetter.

In the next two forms of condensation you will consider, **dew** and **frost**, the water-drops form on objects on the surface of the earth.

67. (L.) Drop some pieces of ice into a flask of water and stand the flask in a warm room. What is formed on the outside of the flask? Where does it come from?

68. (I.L.) Cool some water in a flask by adding ice or ammonium chloride in small quantities at a time. Keep a continuous record of its temperature as it cools. Mark the first temperature at which dew first appears on the outside of the flask.

The last two exercises illustrate the formation of **dew**. At night, soon after sunset, the leaves of trees and plants, blades of grass, and other objects on the surface of the earth become cold like the outside of the flask in the two last exercises. The layer of air next to the earth's surface, like the layer of air near the flask, becomes cold and loses its moisture. This moisture is deposited on the blades of grass etc. just as it is deposited on the outside of the flask.

69. (O.) When there has been a deposit of dew make a list of objects on which the dew has formed, and another list of objects on which the dew has not formed.

70. (G.) What weather conditions are necessary for the formation of dew?

71. (V.) Why does more dew form in valleys than on hill-tops?

72. (G.) Does dew form on a cloudy or a windy night?

The white **hoar-frost** which you know well (x in the notation of Chapter II) is formed in the same way as dew, but instead of water-drops on the blades of grass etc. we should find little ice particles. Perhaps you have noticed how, on a winter's night, the windows of a warm room have been streaming with moisture, and how, in the morning, when the room has become cold, the window panes have been frosted over with beautiful designs of ice particles.

Hoar-frost may be described as a feathery deposit of ice which forms upon leaves, twigs, blades of grass, and other objects on the earth's surface.

EXERCISES.

73. (G.) What is the difference between a "black frost" and a "white frost"?

74. (G.) What weather conditions are favourable for frost?

75. (G.) What particular wind brings more frost than any other wind?

76. (G.) During what part of the year do we get warm days and cold frosty nights?

77. (O.) What is the effect of a sharp frost on growing potato plants in late spring?

78. (O.) After an early autumn frost name the plants which have been killed.

79. (O.) How many days of frost does it take before the ice on a pond is thick enough for skating?

80. (O.) When any damage has been done to plants by frost note the heights above the ground to which the damage extends.

81. (O.) After a winter frost measure the thickness of ice in a number of places exposed and sheltered. Compare the measurements you have made.

82. (E.) During an intense frost pour down some water in any place suitable for making a slide. Describe exactly what you see happen after the water has been poured down. Does the result differ when you pour from a bucket, a garden watering-can with a rose, or spray with a hose-pipe?

83. (E.) When the trees are white with rime describe exactly what you see when the rime is examined with a hand lens.

84. (E.) After a heavy hail-storm collect a dozen of the biggest hail-stones. Examine them carefully and describe how they are made. Put a number of the hail-stones that have not been damaged, nor have been allowed to melt, into a glass beaker. Warm the beaker carefully so as to melt the hail-stones. Measure the resulting water in c.c. and find the average weight of the hail-stones.

CHAPTER V

RAIN, SNOW AND HAIL. THUNDERSTORMS

EXERCISE.

85. (V.) What are the four forms of condensation of water-vapour with which you have already dealt ? Which form occurs in a solid state ? In what form do the others occur, and how do they differ as to where they occur ?

There are seven forms of condensation in all. The three you have not dealt with yet are rain, snow, and hail. These three forms of condensation differ from the previous forms in that the particles of condensation **fall** from the air to the ground.

Raindrops are formed from cloud particles. Some raindrops are very small, others are as large as one-tenth of an inch in diameter.

EXERCISES.

86. (V.) When do you see the largest raindrops ?

87. (V.) What kind of cloud brings the most rain ?

When the air is sufficiently cold, snow-flakes are formed instead of raindrops and instead of a fall of rain you get a fall of snow.

EXERCISES.

88. (G.) Describe the effect of a snow-storm on a countryside.

89. (O.) When snow is falling catch a few flakes on a piece of black cloth. Do not let the flakes melt, but look at them through a magnifying glass. You will find that snow-flakes are of many varied and beautiful shapes.

90. (O.) Whenever there is a fall of hail, examine the hail-stones, and make a note of their size and appearance.

Hail-stones are small opaque pellets of ice, frozen raindrops in fact. Examination of a hail-stone may show that the central ice pellet has had several additional layers of ice frozen on to it during its whirlings aloft in the air. There is another form of hail which is called **soft hail**. The pellets are whiter than the pellets of the other kind of hail, and they are quite soft. This soft hail is most frequent in spring and autumn.

EXERCISE.

91. (G.) During what kind of a storm does hail form in summer time? How do the hail-stones in such a storm compare with winter hail-stones?

RAINFALL.

There are many reasons why rainfall should be placed first in importance amongst the things you are studying in connection with the weather. People are dependent upon rainfall for water supply. Rainfall feeds the rivers and lakes, and without it crops would not grow.

In any country or district it becomes a matter of necessity to have an accurate knowledge of the rainfall throughout the year. With such a knowledge an idea as to the amount of water supply can be formed, and a farmer can tell to some extent what crops he can grow.

So far all you have done with regard to rainfall has been to enter the letter *r* (raining) in your observation book when rain was falling at the time of observation. You will now have to adopt some means of measuring the amount of rainfall.

EXERCISE.

92. (O.) Place a small flat-bottomed tin out in the open on a rainy day. After a time estimate or measure the depth of rain that has fallen into the tin.

Suppose that in Exercise 92 the estimated depth of rain were 2 millimetres. That would mean that, if all the rain had remained in the place where it had fallen, none of it having been lost in any way, there would have been water standing on the ground to a depth of 2 millimetres. Rainfall measurements are always made and given in this way. The amount of rain is given in inches or millimetres, and the measurement gives the actual depth of the water which has fallen on a certain area in a certain time, generally 24 hours.

It would not do to use a small tin to measure rain as in the last exercise, because some of the rain would be lost by evaporation. To prevent loss of collected rain a rain-gauge must be used. All records of rainfall are based on measurements made with rain-gauges. It is quite an easy matter to make a simple rain-gauge.

EXERCISE.

93. (V.) After a shower of rain where does the rain that has fallen on the pavement disappear to as the pavement dries ? To what process is this the opposite ?

EXERCISE.

94. (O.) *To make a rain-gauge.*

Obtain an ordinary tin funnel, the diameter of whose rim is about 14 cms. Obtain also a glass bottle which will hold not less than half-a-gallon of water. Measure as accurately as you can the inside diameter of the funnel rim. Now obtain a glass measuring cylinder

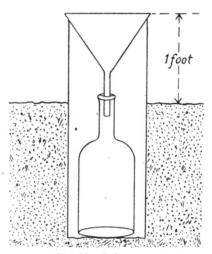

Fig. 30. Rain-gauge.

graduated in cubic centimetres. The bottle and funnel form a simple rain-gauge. The graduated cylinder can be used as a measuring glass.

Choose a position on or near your observation ground in which to plant your rain-gauge. A very good position is in the centre of a grass plot which is not shaded by big trees or buildings. The further away the trees or buildings are, the better the exposure of the rain-gauge.

Where your rain-gauge is to be fixed, make a hole in the ground slightly larger than the bottle. Place the funnel in the neck of the bottle, and then fix the bottle in the hole you have made in the ground. The rim of the funnel should be horizontal, and at a height of one foot above the ground. If you are able to obtain or make a cylindrical vessel into which the bottle could be placed, and on to the top of which the funnel could be fixed, it would improve your rain-gauge (see Fig. 30). You would fix the cylindrical vessel firmly and permanently in the ground.

When rain is falling, the rain which is caught by the funnel runs down into the bottle. Suppose that your rain-gauge is left out for 24 hours, then the rain in the bottle is the amount of rain which has fallen on the circular area of the funnel rim during the 24 hours. You will have to measure this amount.

MEASUREMENT OF RAINFALL.

To measure the rainfall, remove the funnel of your rain-gauge, take out the bottle and pour the collected rain into your measuring glass. Read the volume of the rain to the nearest half cubic centimetre. Now this is the volume of rainfall which has fallen on the circular area of your funnel rim. Suppose, for example, that this volume is $37\frac{1}{2}$ cubic centimetres, and the inside diameter of your funnel rim is 14·2 centimetres and its radius therefore 7·1 centimetres. What you want to know is the depth to which the rain would have stood if it had remained exactly where it had fallen, and none of it had been lost.

The area of the funnel rim is $\frac{22}{7} \times 7\cdot1 \times 7\cdot1$ sq. cms.

$= 158\cdot4$ sq. cms.

The volume of water $\qquad = 37\cdot5$ c.c.

Now if d were the required depth of water you have also that the volume of the water $= 158\cdot4 \times d$ cubic cms. (*i.e.* area of funnel rim $\times d$).

These two volumes are the same, therefore

$$158\cdot4 \times d = 37\cdot5,$$

$$d = \frac{37\cdot5}{158\cdot4}$$

$$= \quad \cdot24 \text{ centimetre}$$

$$= 2\cdot4 \text{ millimetres.}$$

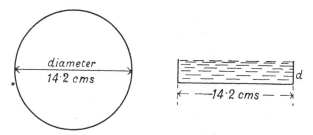

Fig. 31. Measurement of rainfall.

Your rainfall should always be given to the nearest tenth of a millimetre.

Since the area of your funnel rim is the same for all your measurements of rainfall, the depth of rainfall, d, in centimetres is

$$\frac{\text{Measured volume of rainfall in c.c.}}{\text{Area of funnel rim}}.$$

In the above example this is

$$\frac{\text{Measured volume of rainfall in c.c.}}{158\cdot4} \text{ centimetres}$$

$$= \frac{\text{Volume in c.c.}}{158\cdot4} \times 10 \text{ millimetres}$$

$$= \frac{\text{Volume in c.c.}}{15\cdot84} \text{ millimetres.}$$

Hence to get the depth of rainfall in millimetres when the inside diameter of the funnel rim is 14·2 cms., divide the volume of rain measured in cubic centimetres by 15·84. Division by 16 would be sufficiently accurate in this case.

EXERCISES.

95. (I.) From the inside diameter of the funnel rim of your rain-gauge, find the number by which you must divide the measured volume of rain in cubic centimetres in order to get the depth of the rainfall in millimetres.

96. (O.) Every morning at 9 a.m. inspect your rain-gauge and measure the rain (if any) which has fallen during the 24 hours from 9 a.m. the previous day. Enter this amount in millimetres, to the nearest tenth of a millimetre, in your observation book. When you are dealing with daily totals of rainfall the amount you measure one morning at 9 a.m. will be called the rainfall for the previous day. Thus, the rain measured on Thursday morning will be called Wednesday's rainfall.

It would be advisable to make a table for your rain-gauge, such that you could put down straight away from the table the rainfall in millimetres from the reading of your measuring glass in cubic centimetres.

Such a table for a rain-gauge of 14·2 centimetres diameter as used in the example on page 51 would be

TABLE FOR RAIN-GAUGE, DIAMETER OF FUNNEL
RIM 14·2 cms., DIVISION NUMBER 16.

Cubic centi-metres in measuring cylinder	Rainfall in millimetres	Cubic centi-metres in measuring cylinder	Rainfall in millimetres
0·5	Trace	9·0	
1·0		9·5	0·6
1·5	0·1	10·0	
2·0		10·5	0·7
2·5		10·0	0·6
3·0	0·2	20·0	1·2
3·5		30·0	1·9
4·0		40·0	2·5
4·5	0·3	50·0	3·2
5·0		60·0	3·8
5·5		70·0	4·4
6·0		80·0	5·0
6·5	0·4	90·0	5·6
7·0		100·0	6·3
7·5			
8·0	0·5		
8·5			

Example. A reading of 69·5 cubic centimetres is taken

$$60 \quad \text{c.c.} = 3·8$$
$$9·5 \text{ c.c.} = 0·6$$
$$69·5 \text{ c.c.} = 4·4 \text{ millimetres of rain.}$$

EXERCISES.

97. (I.) Construct a table for your rain-gauge similar to that on page 54.

98. (I.) State from your observations which wind brings the most rain in your district.

99. (V.) Have you noticed any relationship between the amount of rainfall and (i) amount of cloud; (ii) height of cloud, (iii) colour of cloud, (iv) type of cloud ?

100. (V.) Has the strength of the wind anything to do with rainfall ?

101. (O.) What happens to rivers and lakes after a lot of rain has fallen ? If possible find out by observation how long it is before the nearest river or stream feels the effect of a few hours or days heavy rain.

102. (E.) Make and fix a gauge or measure by which you can record the height of water in a stream or river daily.

103. (O.) When there has been a fall of snow, measure the depth of the snow somewhere where the fall has been uniform. Enter the result in your note-book. If you cannot find a place where the layer of snow is uniform, take a number of measurements and enter the average in your note-book.

MEASUREMENT OF SNOWFALL.

When snow has fallen into the funnel of your rain-gauge the equivalent amount of rainfall can be got in two ways :

(i) Bring the funnel and bottle indoors into a warm room. Allow the snow to melt, and then measure the water in the usual way.

(ii) Measure out in your measuring cylinder a definite volume of warm water. Pour this into the funnel so that all the snow is melted, and the resultant water runs into the bottle. Then measure the volume of water in the bottle, subtract the volume of warm water you used, and you then have the volume of snowfall in terms of equivalent rain.

When snow is actually falling the second method is the better because it is quicker.

Roughly speaking a foot of snow is equivalent to one inch of rain.

EXERCISE.

104. (O.) After a number of occasions on which snow has fallen compare the depth of snow on the ground as measured in Exercise 103 with the depth of equivalent rainfall measured from your rain-gauge.

THUNDERSTORMS.

EXERCISES.

105. (H.) Describe a typical thunderstorm as fully as you can.

106. (V.) When do thunderstorms occur frequently? In what part of the year are they very rare?

107. (O.) During a thunderstorm sketch the lightning flashes as they appeared to you.

108. (O.) Observe the clouds very carefully during a thunderstorm. Draw sketches of any clouds of peculiar shape.

109. (O.) When a severe thunderstorm is taking place sit with a finger on your pulse. Count pulse-beats between flash and clap and record all observations.

110. (O.) If you happen to have a stopwatch take the time between flash and clap. The distance of the lightning flash can be roughly estimated by counting a mile for every 5 seconds.

111. (O.) Measure the rain that falls during thunderstorms. How does the amount which falls during a severe storm compare with a day's steady rain?

112. (O.) Take frequent wind observations during a thunderstorm. Note carefully any changes in wind direction or strength.

113. (H.) What damage have you seen done by thunderstorms? What does the damage?

114. (V.) What is a lightning conductor? How does it work?

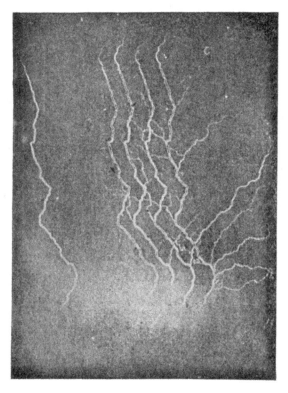

Fig. 32.   Lightning flash taken by a moving camera.

Why is it dangerous for animals to shelter under trees during a thunderstorm?

115. (V.) People sometimes say, "There's thunder about, you can feel it in the air." What is it they can feel? Do you know any signs of thunder?

Whenever thunderstorms occur in your district you should make careful notes of the time, and date. You should also record very carefully the weather conditions before, during and after the storm. If you can get observations from other schools you will find that thunderstorms travel across the country. You may be able with the aid of reports from other schools to track

Fig. 33. Tree on Wandsworth Common, London, struck by lightning during the severe thunderstorm on June 14th, 1914. Seven persons were killed whilst sheltering under the tree, and four others were injured.

a particular thunderstorm quite a long way. By getting the times of occurrence of the storm at various places you may be able to work out its rate of travel. Sketches of the lightning flashes in different storms are worth making. You can make comparisons of the flashes in different storms.

CHAPTER VI

TEMPERATURE AND HUMIDITY

EXERCISES.

116. (V.) When you hear the remark " It is cold this morning" or " It is warm this morning" what is it that feels cold or warm to the speaker ?

117. (V.) Why is it necessary for you to wear an overcoat in winter ? Why does an airman wear a leather coat and thick fur gloves when flying high in summer ?

118. (V.) Which wind would you describe as (i) generally warm, (ii) generally cold?

119. (V.) What is the effect of the springing up of a breeze on a hot summer day ?

120. (V.) Are the coldest days in winter cloudy or clear ?

121. (V.) Is a frosty night generally clear and starry, or is it cloudy and dark ?

122. (V.) In summer are the hottest days sunny or cloudy ?

123. (V.) Is it warmer to-day than it was yesterday ?

124. (V.) Why do you like to bathe in summer and not in winter ?

In these exercises you are discussing a certain kind of change which takes place in the air round about you. You know that there is a layer of air surrounding the earth, and that this layer of air is called the atmosphere. Now in studying the weather you are studying the changes which are continually taking place in the atmosphere. Just as water drawn from a hot water tap feels warm to your hand, and water from a cold water tap feels cold to your hand, so the air round about you sometimes feels cold to you, and at other times feels warm to you. It would be quite easy for you to use the words

hot, warm, cool, cold, etc.

to describe how the air felt to you. Changes from hot to cold, cool to warm etc., are called changes in **temperature**. The words hot, warm, cool, cold, form a rough temperature scale. Such a scale of temperature however would be rather vague, and perhaps misleading at times. You will want a much more accurate scale, and you will therefore have to use an instrument called a **thermometer**. The word thermometer is formed from two Greek words meaning heat-measure.

A thermometer is made from a piece of narrow glass tubing. One end is blown out into a bulb, generally round in shape. The other end is sealed. The bulb and part of the stem of the thermometer contain some fluid, usually mercury. When the temperature increases the fluid in the thermometer rises in the stem. The temperature is given by the position of the top of the fluid on a scale to which the glass tube is firmly attached.

On all temperature scales there are two standard temperatures, the freezing point of water and the boiling point of water. In our country observers of the weather use the Fahrenheit scale of temperature. On this scale the freezing point of water is numbered 32, and the boiling point 212, the intermediate space being divided into 180 equal divisions called degrees. The name Fahrenheit is that of the man who invented this scale. He was born in Dantzig in 1686.

In the science laboratory you use the Centigrade scale. The freezing point of water is numbered 0, and the boiling point 100. There are therefore 100 equal degrees between these two points. This scale was invented by a man named Celsius who was born at Upsala in 1701. Continental observers of weather use the Centigrade thermometer.

EXERCISES.

125. (L.) Examine a Fahrenheit thermometer. Note the temperature of the room. Touch the bulb of the thermometer with your hand and note what happens to the mercury.

126. (L.) Make a Fahrenheit thermometer.

127. (E.) Construct a diagram or table for converting temperatures on the Fahrenheit scale to temperatures on the Centigrade scale and *vice versa*.

128. (I.) What Fahrenheit temperature corresponds to 50° Centigrade, and what C. temp. corresponds to 50° F. ?

Thermometers are used to give the temperature of the air. It is rather difficult however, to obtain a proper exposure for the instrument. If the sun were allowed to shine directly on a thermometer the temperature indicated would be higher than that of the air round the thermometer. The next set of exercises will show you how the position of a thermometer affects the readings it gives.

EXERCISES.

129. (O.) Obtain two similar thermometers. Place them out in the open as near to one another as possible, and so that the sun shines on one while the other is in the shade. Note the differences in temperature shown by the two thermometers at various times.

130. (O.) On a sunny afternoon place one of your two thermometers on the sunny side of a building, and the other on the shady side. Take simultaneous readings of the two thermometers at intervals of an hour throughout the afternoon. Compare the readings.

Does it make any difference if the sun is behind a cloud while you take the readings ?

131. (O.) Place one of your two thermometers about a foot above the ground, and the other about five feet above the ground vertically above the first thermometer. Take simultaneous readings of the two thermometers at various times. Carry out this exercise with both thermometers in sunshine first over a hard smooth surface such as an asphalt pavement, and secondly over short grass.

132. (O.) Repeat the last exercise with the two thermometers in the shade.

You will see that it is very necessary to have some definite plan in taking the temperature of the air, otherwise readings will be obtained which are not comparable with other readings, nor comparable amongst themselves. What observers of weather aim at is to get the temperature of free air in the shade. This is usually done by placing the thermometer inside a shelter or screen, the sides and bottom of which are made of wooden lattice work, the top being of plain wood. The thermometer is thus exposed to a free circulation of air, is protected from direct or reflected sunshine, and is kept dry.

The standard thermometer screen in the British Isles is the Stevenson screen (see Fig. 34). A Stevenson screen is rather hard to make. Full working instructions are given in the *Observer's Handbook of the Meteorological Office, London.*

A good substitute for the Stevenson screen is a ship's screen. This is quite simple in construction. You could perhaps make a serviceable screen of this pattern in your woodwork class. The top is of plain wood, and is made to slope from front to back, the sides and bottom are of simple lattice work (see Fig. 35).

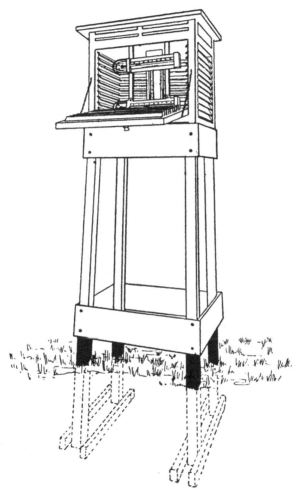

Fig. 34. Stevenson screen with thermometers.

In fixing up a thermometer screen the height should be five feet above the ground. The door should face north so that the sun will not shine on the thermometers when the door is open. The best position for a screen is over short grass.

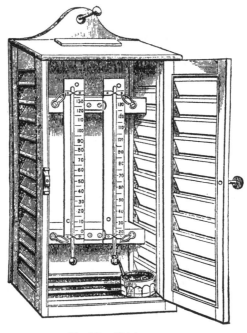

Fig. 35. Ship's screen.

Until you get a screen for your thermometers place one thermometer outside a window on the north side of a building. The thermometer should be placed about a foot from the window in such a position that the instrument can be read without opening the window.

When you read your thermometers you should record the temperature to the nearest half degree Fahrenheit. After a time you should learn to read the temperature to the nearest tenth of a degree.

EXERCISE.

133. (O.) Take the air temperature at your hours of observation. Enter the readings in your note-book.

HUMIDITY.

You have already been told that the air around and above you contains water-vapour.

EXERCISES.

134. (V.) How does this water-vapour get into the air?

135. (V.) By what process does this water-vapour become visible, and in what forms does it appear when visible?

The amount of water-vapour in the air varies from time to time. If there were a lot of water-vapour present the air would feel damp to you. If there were very little water-vapour the air would seem dry. The state of the air as regards the amount of water-vapour present is called its **humidity.** The word humid means damp. When you speak of the humidity of the atmosphere you must understand that you are speaking of the amount of water-vapour present.

There are a number of substances which are sensitive to the humidity of the atmosphere. Seaweed becomes damp and moist when there is a lot of water-vapour in the air, that is when the humidity is high. When the air is dry, or in other words, when the humidity is low, seaweed becomes dry and lighter in colour.

EXERCISES.

136. (O.) Hang a piece of seaweed outside in the shade where it is protected from the rain. Note the changes in its appearance from day to day.

137. (O.) Soak some strips of blotting paper in solutions of (*a*) pure sodium chloride, (*b*) calcium chloride, (*c*) magnesium chloride, (*d*) potassium chloride. Hang these beside the seaweed strip. Report on their behaviour.

138. (O.) Suspend a few fir cones in your house or school where the air can circulate round them freely. Note the kind of weather which causes the cones to close or open their scales.

139. (V.) Have you ever noticed anything unusual about ordinary table salt on a damp day ?

140. (V.) What kind of weather is it that causes the inside walls of a building to stream with moisture ?

The humidity of the atmosphere affects catgut. This fact is made use of in a little toy known as "Jacky and Jenny." In the front of the toy are two openings, out of one of which comes the figure of a woman, and out of the other the figure of a man. When the "old woman" appears it is said to be an indication of fine weather. When the "old man" comes out it is taken to mean bad weather. The two figures are attached to a piece of twisted catgut which winds up or unwinds according as the air gets drier or moister. What this little toy really indicates is whether the air is damp or dry. When the "old woman" is out the air is dry. When the "old man" is out the air is damp.

A human hair is affected by the dampness or humidity of the atmosphere. It increases in length as the air gets moister, and it shortens as the air gets drier.

EXERCISE.

141. (L.) Obtain a long human hair. Wash it carefully in weak ammonia to take out the oil. Fasten one end of the hair to the top

of a small wooden framework, and pass the other end round a small wooden cylinder which will turn very easily. To the loose end of the hair hang a small weight. Attach a pointer to the cylinder and fix a scale in position so that the pointer moves round the scale (see Fig. 36). As the hair lengthens or shortens it will turn the cylinder round. The pointer will turn with the cylinder. You will be able to determine which side of the scale should be marked "damp," and which "dry."

There are certain substances which change colour according to the dampness of the atmosphere. One of these is cobalt chloride.

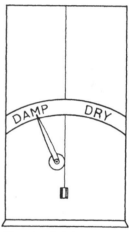

Fig. 36. Hair hygroscope.

EXERCISES.

142. (L.) Obtain some crystals of cobalt chloride. Note the colour. Place in a porcelain dish and heat gently. Note the colour changes very carefully. Allow to cool. Then pass over the dried substance a current of steam, and observe how the substance regains its colour. The steam should be passed over gradually.

143. (O.) Make a strong solution of cobalt chloride and soak a piece of white blotting paper in it. Let the blotting paper dry. Hang up the blotting paper somewhere where the outside air can get to it. Note any colour changes that take place in the blotting paper from day to day according to the dampness of the air.

144. (L. and O.) Record the colour of your cobalt chloride paper (*a*) when placed inside a laboratory desiccator, (*b*) during hard frost, (*c*) during a thunderstorm, (*d*) on the mornings before important cricket matches, football matches, and excursions.

145. (O.) After heavy rain, notice whether pebbles are beginning to dry, or whether pavements remain wet. Do your observations give any clue as to what may happen next?

146. (V.) Have you ever noticed a change in temperature after a sharp shower in summer ? What do you think has caused the change ?

When evaporation takes place over a surface that surface is frequently cooled. You have an example of  this when you fan your face on a hot day. Moisture on your face is evaporated and you feel a cooling effect in consequence. This cooling effect is made use of in order to **measure** the humidity of the air.

EXERCISES.

147. (L.) Use two Fahrenheit thermometers for this experiment. Round the bulb of one of them wrap a piece of muslin. Leave the other thermometer just as it is. Damp the muslin with water. After a little while read the two thermometers. Do their temperatures agree ? What is it that causes one to read lower than the other ?

148. (L.) Repeat the last experiment reading and noting the temperatures shown by the dry bulb thermometer and the wet bulb thermometer, the muslin being damped with (i) distilled water, then (ii) with strong salt solution, then (iii) with methylated spirit, and then (iv) with ether.

Fig. 37.
Wet bulb thermometer.

149. (O.) In your thermometer screen which already contains one thermometer, place a second thermometer. Round the bulb of this second thermometer wrap a piece of muslin. To this muslin attach a piece of cotton wick. Immerse the lower end of the wick in a small vessel containing water. The cotton wick will carry water to the muslin and so keep the muslin damp. Whenever you read the temperature of the air as indicated by your dry bulb thermometer, read the temperature given by the wet bulb thermometer. Enter both readings in your observation book. From the readings of your dry and wet bulb thermometers

describe the state of the atmosphere by using the diagram in Fig. 38. Do this at each of your observation hours, and enter the result in your note-book.

Except when the air is so full of moisture that it cannot take up any more, evaporation goes on round

STATE OF ATMOSPHERE FROM READINGS OF WET AND DRY BULB THERMOMETERS.

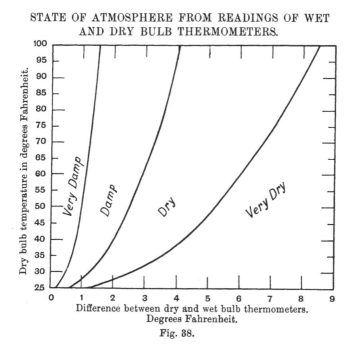

Fig. 38.

the bulb of your wet bulb thermometer. This cools the bulb and the reading of the wet bulb thermometer is lower than that of the dry bulb thermometer. The drier the air gets, the greater is the evaporation, and the greater is the difference between the readings of the dry and wet bulb thermometers.

150. (V.) About what hour of the day would you expect temperature generally to be (i) highest, (ii) lowest ?

151. (V.) How does temperature vary throughout the year ?

152. (H.) When you have taken a whole calendar month's observations of temperature find the average temperature for the month for each of your hours of observation.

153. (H.) Repeat Exercise 152 for each month for a year, and then show by a diagram the average monthly temperatures for a year.

MAXIMUM AND MINIMUM THERMOMETERS.

If you are making a thorough study of temperature at your observation station you will require a maximum and a minimum thermometer.

A maximum thermometer is a thermometer designed to indicate the highest temperature during a certain period of time, *e.g.* a day. It is usually a mercury thermometer, but it has a bore which is constricted near the bulb. When the temperature rises the mercury in the bulb flows past the constriction, but as the temperature decreases the mercury cannot flow back again. It is stopped by the constriction. Thus the mercury stops at the highest temperature indicated. If you read a maximum thermometer always at 9 a.m. it will give you the highest temperature recorded during the previous 24 hours.

The maximum thermometer is set by holding the bulb downwards and then shaking the instrument.

A minimum thermometer is one which records the lowest temperature in a certain period of time. It is usually a spirit thermometer. Inside the tube is a small metal index. As the temperature falls the spirit draws the index back with it. As the temperature rises again, the spirit flows past the index, leaving it where it was. The index thus records the lowest temperature.

This thermometer is set by tilting it with the bulb upwards and allowing the index to run down to the end of the spirit column.

EXERCISES.

154. (O.) If you have maximum and minimum thermometers, obtain the highest and lowest temperatures recorded during the 24 hours ending at 9 a.m. each morning. You will set the two thermometers just after you have read them. When you have set them they should read practically the same as your dry bulb thermometer.

155. (E.) Many plants grow better in a green-house than in a sitting-room. What instruments would you use and what inquiries would you make in carrying out an exact comparison of green-house and sitting-room climate?

156. (E.) Note the temperatures of any green-houses to which you have access.

157. (E.) Note the difference between the dry and wet bulb thermometers at 9 p.m. or 10 p.m. for as many evenings as possible. Note the occurrence of mist or fog. Does the difference between the two thermometers give any indication of mist or fog during the night?

158. (E.) By a comparison of readings of your dry bulb thermometer and the reading of your maximum thermometer state at what time of the day the maximum temperature is reached as a rule.

CHAPTER VII

THE PRESSURE OF THE ATMOSPHERE

EXERCISES.

159. (V.) How do you know that there is air round about you when you are (i) out in the open, (ii) in your class-room ?

160. (V.) How is air necessary to animal and plant life ?

161. (H.) What important constituent of air have you already dealt with in previous chapters ? Write a short account of what you have learnt of it.

162. (H.) Give examples of solids, liquids and gases. In what ways do these three classes of things differ ?

163. (V.) Do you know any of the properties of the gases oxygen and nitrogen ? Discuss any experiments you have performed or seen performed with these gases.

164. (V.) A boy was once asked how he would tell which of the three gases oxygen, hydrogen and nitrogen a jar contained. He replied, " Put a lighted taper in the jar, if it goes off it is hydrogen, if it goes on it is oxygen, if it goes out it is nitrogen." Was his answer correct ?

Air is a mixture of gases. The envelope of air which completely surrounds our earth, and which is called the **atmosphere**, consists largely of two gases, nitrogen and oxygen. Up to a height of at least six miles the greater part of the air is nitrogen. About one-fifth is oxygen. Perhaps you have already learnt in your science work that no fire can be kept going without oxygen, and that neither man nor animal can live without oxygen. There are small quantities of other gases in the atmosphere, and there is of course water-vapour as you have learnt in previous chapters.

Air has weight. Compared with solid bodies such as iron or lead the weight of air is very small. It takes about 13 cubic feet of pure, dry air to weigh 1 lb. A cubic foot of water weighs 1000 ozs., a cubic foot of air weighs $1\frac{1}{4}$ ozs.

EXERCISES.

165. (I.) By calculation find roughly the weight of air in your class-room. How many cubic feet of water would it take to balance it ?

166. (V. Experiment.) Hold your hand out palm upwards. Put a heavy weight in the palm of your hand. Add more weights until you are unable to hold them. Which way is your hand forced to move when you have reached the limit you can hold ? In what direction then do these weights press ? Do you feel pressure in any other direction ?

Air having weight exerts a pressure just as the weights do in the last exercise. To understand exactly how air-pressure behaves you ought to know something about bodies which are called **fluids. A fluid may be described as a substance which flows.** Some fluids like treacle take a long time to flow. Water is a fluid which does not take long to flow. Water flows quickly down hill.

EXERCISES.

167. (V.) What is the name given to substances which do not flow ?

168. (L.) Pick out five fluids and determine by experiment which of the five takes the longest time to flow, and which the shortest.

169. (L.) Invent some experiment for finding the relative rates at which the following liquids flow : ether, alcohol, water, glycerine, treacle.

Air is a fluid because it will flow.   Cold air will flow
down hill.   Air can be made to flow through a pipe.
You can learn from the behaviour of water how air is
likely to behave since both substances are fluids.   You
know that water will find its way through very small
holes.   Air can get through even smaller holes, and it
can get through more quickly than water.   Air comes
into your class-room through small apertures in the
window frames.   It comes under the door, through the

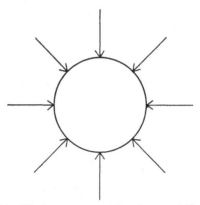

Fig. 39.   The pressure on a sphere immersed in water
acts in all directions.

keyhole, and through the tiniest cracks in the floor.
There is a most important fact to learn here.   **Wherever
air goes it takes its pressure with it.**   Air from outside gets into your class-room, and for that reason the
air-pressure in your class-room is the same as it is
outside.

Suppose a submarine is resting in dry dock, and a
ton weight is placed on the top of it.   This ton weight
exerts a pressure on the submarine in one direction only,

downwards. Suppose that this ton weight has a square base one foot square, then there is a pressure downwards of one ton on one particular square foot of the submarine. Now suppose the submarine is out cruising and submerges to a depth of 40 feet below the surface of the water. The weight of the water above it exerts a pressure on the submarine, not on one particular square foot of it but on every square foot of it. Not only that, but the water exerts its pressure on the submarine in all directions. On the top it exerts a downward pressure, on the bottom of the submarine it exerts an upward pressure, on the sides it exerts an inward pressure. Everywhere it tries to crush in the submarine. Look at Fig. 39 which shows a sphere immersed in water. The arrows indicate the.pressure of the water, **and this pressure acts in all directions.**

Our position on the earth's surface can be compared with that of a submarine resting on the bed of the ocean. The submarine is at the bottom of an ocean of water. We are at the bottom of an ocean of air. The atmosphere is all around and above us just as the water is all around and above the submarine.

EXERCISES.

170. (V.) Why is a fish not squeezed to death by the pressure of the water above it ? Why are we not killed by the air-pressure ?

171. (V.) What happens to fish rapidly brought to the surface from great depths ?

172. (V.) What happens to aviators who fly to great heights ? Why do they carry oxygen with them ?

173. (L.) Obtain an air-tight tin (a 4-gallon petrol tin would do very well). Take out as much air as you can by means of an air-pump. If you can take out enough air the sides of the tin will be crushed in. Why ?

When no air-pump is available you can do this experiment by boiling water in the tin, and, when the tin is full of steam, closing the one outlet you have left, and allowing the tin to cool.

The pressure of the atmosphere is about 15 lbs. per square inch at sea-level.

EXERCISES.

174. (I.) Find by calculation the pressure of the atmosphere on a square foot.

175. (I. Calculation.) What is the pressure of the atmosphere in tons on a square yard ?

176. (V. Measurement and calculation.) What is the pressure of the atmosphere on the floor of your class-room ?

177. (V.) Can you explain why the floor does not give way under so great a pressure ?

178. (V.) Calculate roughly the air-pressure on the back of your hand. Why is it that you can hold your hand out and not feel this pressure ?

179. (H.) Work out the pressure of the atmosphere in the metric system, giving the result in kilograms per square centimetre.

MEASUREMENT OF THE PRESSURE OF THE ATMOSPHERE.

You have not required instruments to enable you to note the changes taking place in many of the weather elements you have considered. You will, however, always require an instrument to measure the pressure of the atmosphere, and to note the changes which take place in it. You can **feel** a change in temperature, but you cannot, in the ordinary way, feel a change in atmospheric pressure. If you were descending the shaft of a coal-pit, or ascending to a great height in an aeroplane you would be conscious of some change or other. You would most likely feel a "singing" in your ears, and very possibly your nose would commence to bleed. Both these things would be due to change of atmospheric pressure, but when you remain on the surface of the earth you do not experience anything which would indicate to you that the pressure of the atmosphere was changing. The instrument which is used to measure the pressure of the atmosphere is called a **barometer**. The first barometer was made in 1643 by an Italian scientist named Torricelli. You can make a barometer just as he did.

Exercise 180. (L.) Obtain a glass tube about three feet long, open at one end, closed at the other. The inside diameter of the tube should be about a quarter of an inch. Fill the tube completely full of mercury. Hold your thumb over the open end of the tube, insert the tube over a small vessel containing mercury. Be very careful not to let any of the mercury fall out of the glass tube. Still holding your thumb firmly over the open end of the glass tube, immerse the open end, thumb as well, in the mercury in the small vessel. When the open end of the tube is well below the surface of the mercury in the small vessel take your thumb away. Clamp the glass tube in a vertical position. What has happened to the mercury in the tube? Why doesn't all the mercury in the tube run out into the small vessel? What do you think is in the space above the mercury at the top of the tube? Clamp an inch scale vertically against the glass tube, one end of the scale being in the mercury in the small vessel. Measure the height of the mercury in the tube above the surface of the mercury in the small vessel. Place a 4 oz. iron weight on the surface of the mercury in the small vessel. What happens to the mercury in the tube?

The glass tube, small vessel, mercury and scale form a simple barometer.

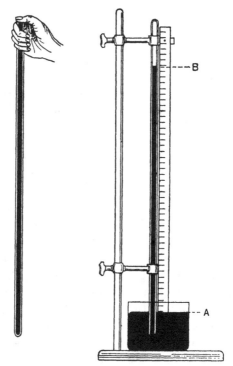

Fig. 40.   Simple barometer.

Exercise 181. (L.) What happens when the barometer tube is inclined? What happens if you again put your thumb under the open end, hold tight, invert the tube, and try to make the "vacuum" run up and down the tube?

Suppose the reading on the scale at B is 31·2 inches, and the reading at A is 1·2 inches. By subtraction you have that the length of the vertical column of mercury from A to B is 30 inches. Now suppose that the area of the inside circular section of your glass tube is "a" square inches. Then the number of cubic inches of mercury in the tube from A to B is $a \times 30$ [volume of a cylinder=area of base $\times$ height]. A cubic inch of mercury weighs roughly $\frac{1}{2}$ lb.

Therefore

$30 \times a$ cubic inches weigh $30 \times a \times \frac{1}{2}$ lb. $= 15 \times a$ lbs.

Now this is the pressure of the mercury on "a" square inches, *i.e.* on the area of the inside section of the glass tube. By simple proportion you have that the pressure on one square inch is 15 lbs. Now this mercury pressure is balanced by the pressure of the atmosphere, hence the pressure of the atmosphere is in this case 15 lbs. per square inch.

As a rough approximation you will see that the pressure of the atmosphere in lbs. per square inch can be got by dividing the height in inches of your column of mercury (A to B in Fig. 40) by 2.

Other liquids could be used to make barometers. A water barometer was once made. Its tube was 36 feet long. A glycerine barometer would be about 30 feet in height.

The barometer in general use in the British Isles is the Kew Pattern Station Barometer. A photograph of an instrument of this kind is reproduced in Fig. 41.

Another type of barometer working on quite a different principle is the aneroid barometer. In this instrument there is a circular metal box which is

Fig. 41.   Station barometer as mounted at South Farnborough.

air-tight, and nearly exhausted of air. The ends of the box are affected by the pressure of the atmosphere. When the pressure increases the ends of the box are pressed inwards. When the pressure decreases the ends of the box move outwards again. The movements of the ends of the box are made to work a pointer on the face of the instrument. The position of the pointer on a circular scale gives the pressure of the atmosphere.

An aneroid barometer is small and it is easily carried, but it is not so accurate as a mercury barometer.

PRESSURE UNITS.

You have seen that the pressure of the atmosphere can be given as so many lbs. per square inch. At sea-level the pressure varies from about $13\frac{1}{2}$ to $15\frac{1}{2}$ lbs. per square inch.

EXERCISE.

182. (H.) Calculate the range of atmospheric pressure at sea-level in (i) cwts. per square foot, (ii) tons per square yard.

If a man closes his fist and holds it with the back of his hand upwards the pressure of the atmosphere on the back of his hand is about 1 cwt. The air pressure on the top of your head may be as much as a quarter of a ton.

The study of the atmosphere and the changes which take place in it is a scientific subject. In fact it has been called the Physics of the Atmosphere. You know that in Physics, or in any science subject, the units used are those of the Metric System, the gram and the centimetre. You ought therefore to use metric units in your study of the atmosphere. Instead of using a unit

C.   6 of pressure of one lb. per square inch, you should use a unit called the **millibar** which is the unit of atmospheric pressure in the metric system. This unit of pressure, the millibar, differs very little from a pressure of one gram per square centimetre. It is easy to remember that a millibar is about one-thousandth part of the standard atmospheric pressure at sea-level.

You have been taking observations of wind on the Beaufort Scale, and you are able to give the force of the wind according to that scale. You know a wind of force 6. It is a wind in which an umbrella could only be used with difficulty. A pressure of one millibar is almost the same as the pressure due to a wind of force 6. When you think that the pressure of the atmosphere is over a thousand times greater than this you will see how very great the pressure of the atmosphere really is.

EXERCISES.

183. (H.) From the following data draw a diagram by which you can convert inches of height in a mercury barometer to millibars and *vice versa* :

BRITISH ISLES.

Normal pressure		1013·2 millibars or	29·92	mercury inches
Highest recorded pressure	1053·5	,,	31·1	,,	,,
Lowest ,, ,,	925·5	,,	27·33	,,	,,

184. (E.) Continental observers of weather have their barometers graduated in millimetres. Given that 10 inches = 254 millimetres draw a graph by which you can change millimetres to inches and *vice versa*. The range of your diagram should be from 27·3 to 31·2 inches.

185. (E.) Using the data given in the last two exercises draw a graph for the conversion of millimetres to inches and *vice versa*.

If you are able to purchase a barometer as illustrated in Fig. 41, graduated in millibars, you should do so and read it at each of your observation hours. If you cannot purchase such a barometer use the simple barometer made in Exercise 180. Write down the height as read in inches, divide by two, and thus get the approximate pressure in lbs. per square inch. Enter these values in your note-book.

EXERCISE.

186. (L.) *To read a Kew pattern barometer graduated in millibars.*

Your Kew pattern barometer will have been mounted ready for reading as shown in Fig. 41. On the right-hand side of the instrument just below the scale you will find a screw-head. Turn this to the right and you will see that it moves two pieces of metal up inside the glass. One of these pieces of metal, called the **vernier**, is at the front of the mercury tube. You will see that its left edge fits close up to the scale of the instrument. The second piece of metal is at the back of the mercury tube. These two sliding pieces of metal are fastened rigidly together. Their lower ends are on the same horizontal level. Turn the screw-head to the left and you will find that the two sliding pieces move downwards.

To set the vernier (sliding pieces) for a reading turn the screw-head until you get the uppermost part of the curved top of the mercury column on the horizontal level of the lower ends of the two sliding pieces. You will have to shut one eye to do this, and you must be very careful to get your open eye at the same horizontal level as the lower ends of the sliding pieces. When the vernier is properly set first read the value of the scale division next below the zero mark on the vernier. In Fig. 42 this is 1012 millibars. Now look along the

6—2

vernier until you see one of its divisions which coincides exactly with a millibar scale division. This in Fig. 42 is 7. The barometer reading is then written down as 1012·7 millibars. If the zero mark on the vernier coincides exactly with a millibar scale division, the 10 mark on the vernier will also correspond with a millibar scale division.

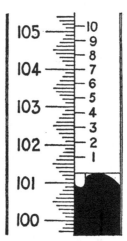

Fig. 42.   Scale and vernier
on barometer.

OBSERVATION BOOK.

Your observation book can now be ruled off as in this example. The order of the observations is the order in which they are taken in this book. There are no more observations to be added. Enter your barometer readings opposite "as read." You will learn what "correction" and "M.S.L." mean in the next chapter.

Date and day of week	Hour of obs.	Wind		Weather		Cloud		Rain-fall in mm.	Temperature				Barometer	Description of clouds Remarks
		Direc-tion	Force	At time	Since last obs.	Types	Direc-tion		Dry	Wet	Max.	Min.		
	9 a.m.												As read Correction M.S.L.	
	9 p.m.												As read Correction M.S.L.	

CHAPTER VIII

WEATHER CHARTS

[Apparatus—Blank maps of N.W. Europe. Copies of the Daily Weather Report from the Meteorological Office, London, S.W.]

You will have noticed, even from a few barometer readings, that the pressure of the atmosphere is continually changing.

EXERCISES.

187. (V.) What kind of weather is usually associated with a low barometer? When the barometer is high what kind of weather do you expect?

188. (I.) Draw a diagram showing the inscription on your weather-glass at home. Is the inscription an accurate one?

If you are only able to obtain readings from your own barometer, you cannot proceed very far with the study of the connection between changes in atmospheric pressure and weather. The readings of your barometer and thermometer, the winds you observe, and the weather you are experiencing are only just a small part of a whole series of events taking place over many thousands of square miles. To know what is really going on you must be able to study a set of observations similar to your own made all at the same time by a large number of observers situated in your own and neighbouring

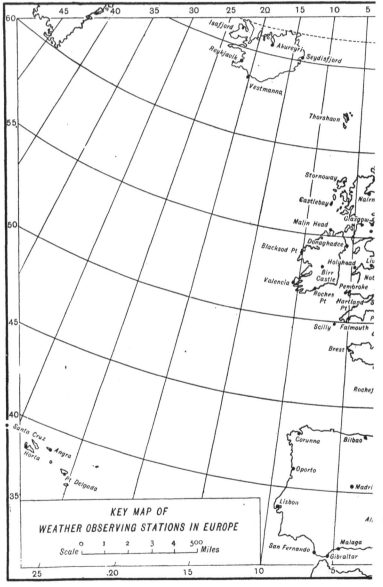

Fig. 43. Weathe

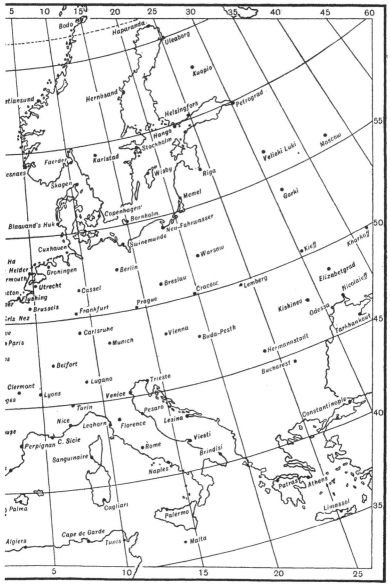

stations in Europe.

countries. You can do this best by making use of the facilities offered you by the Meteorological Office in London. This office directs and controls the meteorological work of the British Isles. It has observers at many places in these islands, and it obtains information from weather observing stations in Europe. There are over 150 weather observing stations reporting to the Meteorological Office in London. Thirty of these stations are in our own islands, mostly on the coast. Outside our islands there are observing stations in Iceland, at the Azores, and on the continent of Europe.

At seven o'clock every morning (Greenwich or Winter time) the observer at each of these stations reads his barometer and thermometers. He notes the direction and strength of the wind, the state of the sky, and the weather. He measures the amount of rain which has fallen since the previous evening or morning. Then he telegraphs his observations to London. As the telegrams are received at the Meteorological Office, the official on duty plots the observations on a large working chart of Europe. At the position of each observing station on the map he draws an arrow showing the direction of the wind at that station. On the arrow he puts flèches according to the force of the wind just as you have them in the table of Beaufort numbers for wind force on page 13. Then he writes close to the position of the station on the blank map the reading of the barometer at that station. Directly under this barometer reading he writes the temperature, and directly under the temperature reading he writes "present weather" in Beaufort notation. Sometimes he puts down "past weather," using red ink to distinguish it from "present weather." He also writes down what the barometer has been doing during the three hours previous to the observation at each station. When this has been done for as many observing stations as possible he completes his weather map. To understand what this weather map really is, we shall consider each thing separately, starting with the barometer readings.

EXERCISES.

189. (V.) Two submarines are submerged in the sea. One is 10 feet, the other 100 feet below the surface. Which one is subjected to the greater water-pressure, and why?

190. (V.) When you have answered the last question try to answer this one: Two exactly similar barometers are placed one at the top, the other at the foot of a mountain. Which one would show the greater pressure, and why?

191. (O.) Find out by experiment (*e.g.* in a church tower or high lift) how many feet you must carry an aneroid barometer up vertically for the pressure of the atmosphere to be reduced by one millibar.

192. (O.) Read an aneroid barometer in your class-room. Note the reading. Carry the aneroid to the top of a hill or to the top of the highest building you have access to. Read the instrument again. Compare the two readings. Repeat the experiment a number of times for hills or buildings of different heights. Get the heights of the hills or buildings as accurately as possible, and see if you can draw a diagram showing the relationship between the height you ascend and the difference that height makes in the pressure reading.

193. (V.) How would the reading of an aneroid be affected by taking the instrument up in an aeroplane?

You will have come to the conclusion from these exercises that the reading of a barometer depends to some extent on the height of the instrument above sea-level. A barometer at the top of a tower 450 feet high would always read about 12 millibars below a barometer at the foot of the tower. It is necessary to make allowance for height above sea-level when a set of observations from various observation stations is being considered. At the telegraphic observing stations of the Meteorological Office, London, the observer makes this allowance before telegraphing his reading. He adds a certain amount to the actual reading of his instrument. This amount depends on his height above mean sea-level (M. S. L.). To use the proper phrase **he reduces his reading to mean sea-level**. Thus all the readings received at the Meteorological Office are comparable one with another, all being given just as if the barometers had been read at sea-level. These corrected readings are the readings which the official on duty at the Meteorological Office puts on his map.

If you have a Kew Pattern Station Barometer you can get a correction card from the Meteorological Office, London, by sending your height above sea-level. This card will tell you the proper number of millibars to add on to a reading of your barometer. In your observation book you should enter the reading of your barometer under "as read." You should then get the proper correction from your correction card, put this down under "correction," add the correction to the reading, and so get the M.S.L. reading.

EXERCISE.

194. (I.) Obtain a blank map of North-Western Europe. An ordinary outline map will do, but blank copies of the Daily Weather Report can be obtained from the Meteorological Office, London, if thought advisable. Enter on this map the following barometer readings taken at 7 a.m. Greenwich time (8 a.m. summer time) on August 28th, 1917. Write each reading near to the position of the

station to which it refers. You can get the positions of the stations from the key-map in Fig. 43.

Station	Millibars	Station	Millibars
Lerwick	990·0	Nairn	980·0
Deerness	985·8	Aberdeen...	979·0
Stornoway	983·8	Leith	974·1
Castlebay	984·5	Tynemouth	969·8
Glasgow	977·0	Spurn Head	968·6
Eskdalemuir	970·3	Yarmouth	974·3
Malin Head	980·5	Clacton-on-Sea ...	974·8
Blacksod Point ...	985·2	Nottingham	967·9
Valencia	988·4	Benson	971·7
Roche's Point... ...	984·9	South Farnborough	973·1
Birr Castle	981·5	London (Kew) ...	973·4
Donaghadee	976·4	Skudesnaes	991·5
Liverpool	968·7	Faerder	998·7
Holyhead	972·1	The Scaw	997·4
Pembroke	977·5	Blaavands Huk ...	989·4
Hartland Point ...	981·2	Copenhagen	998·7
Scilly	987·0	Bornholm	1000·4
Falmouth	984·5	The Helder	982·8
Jersey	985·8	Rochefort	1001·3
Portland Bill	974·3	Biarritz	1006·0
Dungeness	975·0	Paris	993·2
Dover	978·1	Belfort	1002·0
Wick	983·9	Corunna	1009·0

The pressure at Nottingham is 967·9 millibars, the pressure at Holyhead is 972·1 millibars. What is the pressure at a point half-way between these two stations? Put a small cross at this point. You could write the pressure at Nottingham as (970 − 2·1) millibars, and the pressure at Holyhead (970 + 2·1) millibars. This should help you to see why a point *half-way* between Nottingham and Holyhead was marked. The pressure at Tynemouth is 969·8 millibars and the pressure at Eskdalemuir in Dumfriesshire is 970·3 millibars. Suppose a line was drawn on the map from Eskdalemuir to Tynemouth, and this line was divided into five equal parts, you could find the point

Eskdalemuir					Tynemouth
970·3	970·2	970·1	970·0	969·9	969·8

on this line at which the pressure was 970·0 millibars. Do not draw such a line on your map, but by eye judgment mark a point between Tynemouth and Eskdalemuir where the pressure is 970·0 millibars. You will see that from your mark to Tynemouth should measure 2, while from your mark to Eskdalemuir should measure 3. Now find a point between Spurn Head and Yarmouth where the pressure is 970 millibars. You can do it in this way. Mark the place with a cross.

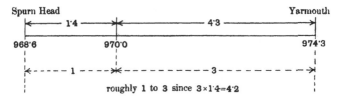

roughly 1 to 3 since 3×1·4=4·2

From this cross to Yarmouth should be three times the distance from the cross to Spurn Head. Find a point between Nottingham and Benson where the pressure is 970·0 millibars. Work in this, a different way.

Benson pressure − Nottingham pressure = 971·7 − 967·9 = 3·8 millibars

970·0 − Nottingham pressure = 970·0 − 967·9 = 2·1　　,,

$\frac{2·1}{3·8}$ is $\frac{1}{2}$ roughly, so mark a point with a cross half-way between Nottingham and Benson. You have now found four places where the pressure is 970 millibars. You have marked these places with small crosses. Find any other places in or near England where the pressure is 970 millibars. Run a line through all the crosses marking places where the pressure is 970 millibars. Compare this line with the innermost line of Fig. 44. Can you find any other place on your map not on the line you have just drawn where the pressure is 970 millibars ?

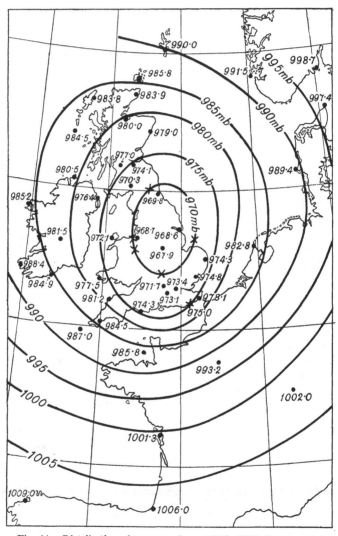

Fig. 44. Distribution of pressure, August 28th, 1917, 7 a.m. G.M.T.

In the last exercise the line you have drawn is called an **isobar**. Along that line the pressure is 970 millibars. An isobar is a line of equal pressure on a weather chart. The isobar you have drawn in Exercise 194 is the 970 millibar isobar.

EXERCISES.

195. (V.) What is a contour line on a map ? Do isobars resemble contour lines ?

196. (I.) On the map you have used in Exercise 194 draw a 975 millibar isobar. Do it by marking with small crosses places on the map where the pressure is 975 millibars. You will mark Dungeness first, the pressure there being 975·0 millibars. Your isobar will run near to Clacton, Yarmouth, Leith, and Portland Bill.

197. (I.) On the map you have used in Exercises 194 and 196 draw 980, 985, 990, 995, 1000 and 1005 millibar isobars. Check your diagram by comparison with Fig. 44.

198. (V.) In Figure 44 isobars are drawn over the North Sea. There are no barometers there. How is it known that the pressure over the North Sea varied from about 970 to 990 millibars for this weather chart ?

199. (I.) On a blank map of North-West Europe, similar to the one you have used in Exercise 194, put the wind readings (given on next page) which refer to August 28th, 1917, 7 a.m. Greenwich time, the same time for which barometer readings were given in Exercise 194.

200. (V.) At what two places in England are the winds exactly opposite in direction on this wind chart ? Over what part of the British Isles are the winds (i) Westerly, (ii) Southerly, (iii) North-Easterly, (iv) Northerly ? Over what country are the winds South-Easterly ?

201. (I.) Suppose that a balloon had been liberated at Yarmouth at 7 a.m. August 28th, 1917, and that it had been blown along by the wind, always being kept near the ground or sea. At first it would have been blown by the strong Southerly wind in a Northerly direction out into the North Sea. There it would have struck the S.E. wind between Blaavands Huk and Tynemouth. This wind would have blown it towards Aberdeen say. Before it had reached the Scottish

coasts it would have been travelling Westwards on the Easterly winds. Then it would have been blown towards Ireland by the North-Easterly winds over Southern Scotland. Suppose that the balloon

Station	Wind direction and Beaufort force	Station	Wind direction and Beaufort force
Lerwick	E 4	Nairn	ENE 3
Deerness...	E 5	Aberdeen	ENE 5
Stornoway	NE 7	Leith	NE 7
Castlebay	N 4	Tynemouth	SE 3
Glasgow	NNE 3	Spurn Head	SSE 5
Eskdalemuir	NNE 8	Yarmouth	S 7
Malin Head	N 6	Clacton-on-Sea ...	SSW 8
Blacksod Point ...	NE 5	Nottingham	S 2
Valencia	WNW 4	Benson	SSW 5
Roche's Point ...	NW 4	South Farnborough	SW 7
Birr Castle	W 1	London (Kew) ...	SSW 6
Donaghadee	NE 5	Skudesnaes	ESE 4
Liverpool	SW 1	Faerder	E 5
Holyhead	NNW 8	The Scaw	SE· 2
Pembroke	W 6	Blaavands Huk ...	SE 7
Hartland Point ...	W 7	Copenhagen	SE 3
Scilly	WNW 7	Bornholm	SW 3
Falmouth	W 7	The Helder	S 8
Jersey	W 8	Rochefort	SW 7
Portland Bill	W 9	Biarritz	WNW 4
Dungeness	SW 10	Paris	SSW 6
Dover	SW 10	Belfort	W 3
Wick	E 4	Corunna...	SW 5

had reached the East of Ireland. It would have been blown along by a North-Westerly wind and then by a Westerly wind. If it had reached Wales or South-West England the Westerly winds there would have

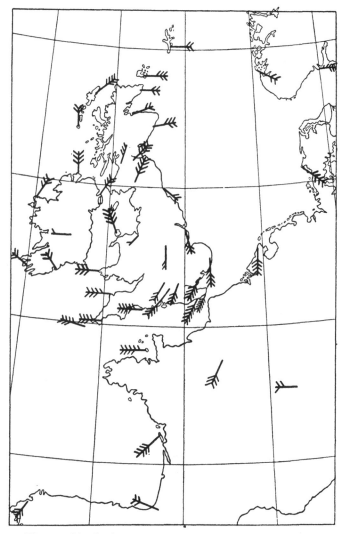

Fig. 45. Distribution of wind, August 28th, 1917, 7 a.m. G.M.T.

blown it towards the centre of England. It might then have come under the influence of the strong South-Westerly winds over S.E. England, and have been blown back to its starting-point at Yarmouth. Supposing all this had been possible draw the track of the balloon on a blank map, and compare this track with the run of the isobars in Fig. 44 (or the map you have used in Exercise 197). What do you notice ?

202. (V.) Put the two charts you have used in Exercises 194 and 199 before you and answer these questions :

(i) A coastguard stands with his back to the wind at Holyhead at the time these charts are drawn for. Is the barometer lower to his left hand than to his right ?

(ii) Is the same thing true at Spurn Head, Dungeness, and Lerwick ?

(iii) Can you find any place where it is not true ?

203. (V.) Suppose that the balloon in Exercise 201 had been big enough to carry a man, and that this man had continually looked in the direction he was travelling (his back would have been to the wind all the time), where would the low barometer have been with respect to him all the time ?

204. (V.) Again keeping the charts drawn in Exercises 194 and 199 before you, answer these questions :

(i) Where are the winds strongest ?

(ii) Where are the isobars closest together ?

(iii) What is the wind force at Dungeness ? What is the distance in miles between the 975 and 980 millibar isobars there ?

(iv) What is the distance between the same two isobars at Aberdeen, and what is the wind force there ?

205. (V.) For the charts used in the last exercise compare the strength of the wind and the nearness of the isobars for as many places as possible. What conclusion do you come to ?

206. (I.) Trace or sketch the isobars from your chart drawn in Exercise 194 (see Fig. 44) on to the chart you have drawn in Exercise 199 (see Fig. 45). You have then a chart showing isobars and winds.

In 1857 Professor Buys Ballot of Utrecht stated this law for the Northern Hemisphere :

Stand with your back to the wind and pressure is lower on your left hand than on your right.

This law is a very important one. You should always keep it in mind when drawing isobars.

EXERCISES.

207. (V.) What exercises have you had already which illustrate Buys Ballot's Law ?

208. (V.) Referring to your chart of Exercise 206 state whether pressure is lowest to the east or west of Nottingham. Give a reason for your answer.

209. (I.) From the set of observations for September 10th, 1917 (see p. 98), put barometer readings and winds on a blank map. Draw a 1025 millibar isobar, then a 1020, then a 1015, and then a 1010 millibar isobar. Draw also a 1024 and a 1023 millibar isobar, but draw these in broken lines. Compare your isobars with those of Fig. 46.

210. (V.) Can you track out the course of a balloon on this chart similar to the one you drew in Exercise 201 ?

211. (V.) Repeat Exercise 202 for this chart.

212. (V.) With the chart of Exercise 209 before you answer these questions :

(i) Where are the winds strongest ?

(ii) Where are the isobars closest together ?

(iii) What is the wind force at Skagen ? What is the distance in miles between the 1020 and 1015 isobars near Skagen ?

(iv) What is the wind force at London ? What is the distance in miles between the 1025 and 1020 isobars through London ?

(v) Is the conclusion you came to in Exercise 205 as to force of wind and nearness of isobars strengthened ?

(vi) Is Buys Ballot's Law true for this chart ?

(vii) Is the highest pressure on the chart to the North or South of Yarmouth ? Why ?

C.   7

Observations for 7 a.m., September 10th, 1917.

Station	Barometer millibars	Wind	Station	Barometer millibars	Wind
Lerwick...	1018·6	W 3	Leith	1024·0	W 1
Deerness	1021·0	WSW 4	Tynemouth	1024·9	SW 2
Stornoway	1021·6	S 2	Spurn Head... ...	1026·2	SE 1
Castlebay	1023·5	SW 4	Yarmouth	1025·1	ESE 3
Glasgow	1024·2	ESE 1	Clacton-on-Sea ...	1025·4	E 3
Eskdalemuir... ...	1025·0	Calm	Nottingham	1025·5	E 1
Malin Head	1022·7	E 2	Benson	1023·6	E 2
Blacksod Point ...	1022·8	E 2	South Farnborough	1023·8	E 2
Valencia	1022·1	Calm	London (Kew) ...	1024·7	E 2
Roche's Point ...	1021·9	NNW 1	The Scaw (Skagen)	1015·5	WNW 5
Birr Castle	1022·6	N 1	Blaavands Huk ...	1022·2	NW 4
Donaghadee... ...	1023·0	NNE 2	Copenhagen	1017·8	WNW 4
Liverpool	1024·1	ESE 3	Bornholm	1016·6	WNW 6
Holyhead	1022·8	ESE 2	The Helder	1026·0	NE 2
Pembroke	1022·8	SSE 3	C. Gris Nez	1024·7	ESE 3
Hartland Point ...	1022·8	NE 2	La Hève	1021·8	E 4
Scilly	1021·7	Calm	Brest	1021·6	E 3
Falmouth	1022·1	ENE 2	Rochefort	1020·0	ENE 1
Jersey	1021·1	E 3	Biarritz	1016·7	E 2
Portland Bill ...	1023·1	ENE 4	Paris	1023·1	NE 1
Dungeness	1023·6	ESE 3	Belfort	1024·2	ENE 2
Dover	1024·7	SE 3	Lyons	1021·4	Calm
Wick	1022·2	WSW 2	Nice	1018·4	Calm
Nairn	1022·8	Calm	Perpignan	1019·8	SW 1
Aberdeen	1023·7	WNW 2	Corunna	1017·2	ESE 1

Just as you have isobars, lines of equal atmospheric pressure, so you can have **isotherms**, lines of equal temperature.

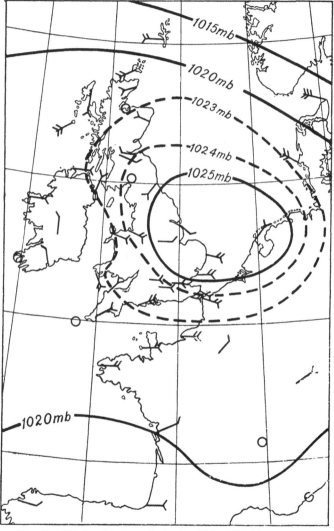

Fig 46. Wind and isobars, September 10th, 1917, 7 a.m. G.M.T.

7—2

EXERCISE.

213. (I.) Enter the following temperatures on a blank chart.

Temperatures at 7 a.m., Greenwich Time, August 28th, 1917.

Station	Temperature Fahrenheit	Station	Temperature Fahrenheit
Lerwick	58	Nairn	55
Deerness	55	Aberdeen	55
Stornoway	55	Leith	55
Castlebay	55	Tynemouth	55
Glasgow	52	Spurn Head	56
Eskdalemuir	52	Yarmouth	59
Malin Head	57	Clacton-on-Sea ...	59
Blacksod Point ...	54	Nottingham	55
Valencia	56	Benson	55
Roche's Point ...	54	South Farnborough	56
Birr Castle	53	London (Kew) ... 57	57
Donaghadee	55	Skudesnaes	61
Liverpool	53	Faerder	61
Holyhead	54	The Scaw	61
Pembroke	55	Blaavands Huk ...	55
Hartland Point ...	56	Copenhagen	59
Scilly	56	Bornholm	63
Falmouth	54	The Helder	61
Jersey	58	Rochefort	63
Portland Bill	58	Biarritz	63
Dungeness	58	Paris	61
Dover	57	Belfort	62
Wick	55	Corunna...	61

Draw isotherms for 60° F. and 55° F. Compare your isotherms with those of Fig. 47.

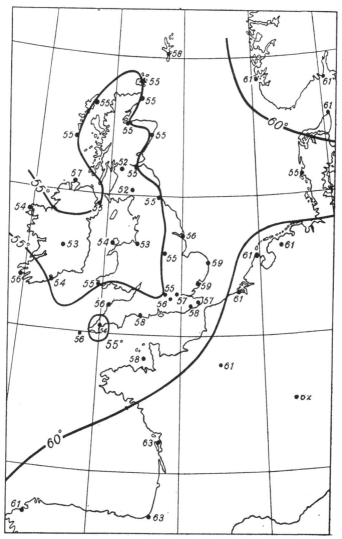

Fig. 47.   Distribution of temperature, August 28th, 1917, 7 a.m. G.M.T.

CHAPTER IX

CYCLONES AND ANTICYCLONES

[Apparatus—Specimens of Daily Weather Report of the Meteorological Office for typical days, 30 copies for each day. Copies of " The Weather Map" by Sir Napier Shaw, from the Meteorological Office, London.]

You have seen in the last chapter how isobars are drawn on a weather chart. You have verified Buys Ballot's Law, and you have seen how the closeness of isobars is connected with the strength of the wind. You will now be able to learn a little of what is known as to the connection between weather and the distribution of pressure as shown by the isobars on a weather chart.

EXERCISES.

214. (I.) On a blank map of N.W. Europe similar to those you have been using in the last chapter enter the following observations for present" weather at 7 a.m. Greenwich Time, August 28th, 1917. Also enter "past" weather during the previous night, 6 p.m. August 27th to 7 a.m. August 28th. For present weather use the following signs :

○	Clear sky	} *b*	≡°	Mist	*m*
◑	One-quarter clouded		≡	Fog	*f*
◐	Half clouded ... ... *bc*		Ⓚ	Thunderstorm	*tlr*
◕	Three-quarters clouded *c*		✳	Snow	*s*
◕	Overcast ... ... *o*		▲	Hail	*h*
●	Rain ... ... *r*				

Enter "past" weather in Beaufort letters. Compare your work with Fig. 48.

Station	Weather 7 a.m. Aug. 28, 1917	Previous night	Station	Weather 7 a.m. Aug. 28, 1917	Previous night
Lerwick...	*om*	*omr*	Nairn	*og*	*orq*
Deerness	*o*	*rbo*	Aberdeen	*r*	*bc, b, or*
Stornoway	*opq*	*opq*	Leith	*r*	*bc, or*
Castlebay	*bc*	*bc, c, p*	Tynemouth	*oq*	*omqr*
Glasgow	*op*	*c, ogp*	Spurn Head...	*orq*	*orq*
Eskdalemuir ...	*or*	*o, or*	Yarmouth	*bcq*	*bcqor*
Malin Head	*cq*	*cq*	Clacton-on-Sea ...	*r*	*ogr*
Blacksod Point ...	*bcqp*	*omqp*	Nottingham	*o*	*or*
Valencia	*o*	*or*	Benson	*r*	*r*
Roche's Point ...	*o*	*ogro*	South Farnborough	*or*	*or, c*
Birr Castle	*r*	*or*	London (Kew) :...	*ouq*	*oruq*
Donaghadee	*ogmq*	*ogm*	Skudesnaes	*o*	*or*
Liverpool	*o*	*or*	Faerder	*o*	*o*
Holyhead	*orqm*	*orqm*	The Scaw	*o*	*or*
Pembroke	*cqpr*	*cqp*	Blaavands Huk ...	*r*	*bc, or*
Hartland Point ...	*r*	*c, oqr*	Copenhagen	*o*	*o*
Scilly	*cqp*	*cqp*	Bornholm	*m*	*r, m*
Falmouth	*o*	*drq, oq*	The Helder	*bc*	*c, bc*
Jersey	*oqp*	*oqr*	Rochefort	*o*	*o*
Portland Bill ...	*cq*	*bc, cqr*	Biarritz	*r*	*or*
Dungeness	*o*	*qp*	Paris	*bc*	*bc, co*
Dover	*r*	*rq*	Belfort	*o*	*o*
Wick	*om*	*orq, op*	Corunna	*bc*	*co*

215. (H.) Describe in words the weather over the British Isles as it was at 7 a.m. August 28th, 1917. Describe also very briefly the weather of the previous night over England, and over France.

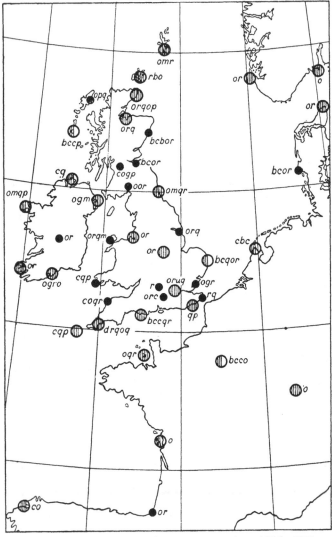

Fig. 48. Distribution of weather at 7 a.m. on August 28th, 1917, and weather previous night.

The pressure distribution at 7 a.m. on August 28th, 1917, is an example of a well-defined type of pressure distribution known as a **cyclonic depression**, or a **cyclone**, or a **depression**, or more simply "a low." In this type of pressure distribution the lowest barometer reading is in the centre, and the pressure increases in all directions outwards from the centre. Put before you the chart you have drawn in Exercise 194. You will notice that the 970 millibar isobar has inside it all the pressures that are below 970 millibars. All pressures above 970, but below 975 millibars lie between the 970 and 975 millibar isobars, and so on for other pressures, the pressures increasing in every direction outwards from the centre of the depression.

From Exercise 201 you know that there is a complete circulation of wind round the depression, the winds going round in a direction contrary to the hands of a watch. Everywhere, in agreement with Buys Ballot's Law the low pressure is on your left-hand side if you stand with your back to the wind.

You must study this type of pressure distribution very carefully. Refer now to Fig. 49, which embodies Figs. 44, 45, 47 and 48. The barometer readings are not written in, but the isobars are shown exactly as in Fig. 44. The winds are shown as in Fig. 45. The temperatures are shown as in Fig. 47, and the isotherms are indicated by broken lines. The "present" and "past" weathers are given just as in Fig. 48.

Fig. 49 is a complete weather chart for 7 a.m. August 28th, 1917. Look at this weather chart carefully. Note that rain is shown to have fallen at most stations in the British Isles during the night previous to the morning on which the observations were taken. Note that it is raining at Blaavands Huk, and that the sky is clearing

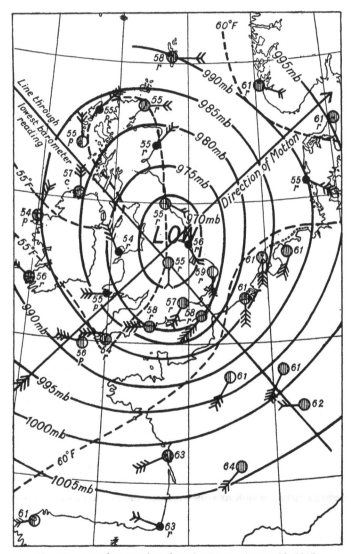

Fig. 49. Complete weather chart for 7 a.m., August 28, 1917.

in the North-West of Ireland, and at Castlebay, although there have been showers in the night in these districts. Note also that the isotherms show that it was warmer to the East of the cyclone than to the West of it.

There are many distinctive features about the distribution of weather in a cyclonic depression, but perhaps the most important thing to study about such a depression is the way it moves. The whole system of isobars as shown on the weather chart moves bodily forward, taking its distribution of wind and weather with it. Sometimes a series of successive weather charts show a depression to move quickly across our islands from West to East. Sometimes a depression moves slowly, and sometimes it lingers about, and even remains stationary for a while.

The cyclone you have been studying in Fig. 49 could be seen to be approaching the Western coast of Ireland on the weather chart of 7 a.m. August 27th (see Fig. 50). The barometer was then falling quickly at Blacksod Point and Valencia, a sure indication of the approach of a depression. On the next chart, 6 p.m. Greenwich time, the depression was off the mouth of the Bristol Channel. On the morning of the 28th the depression was centred over the North of England. The next charts (see Fig. 50) show that the depression moved away Northwards, and finally North-North-Eastwards to the North of Scandinavia.

EXERCISES.

216. (I.) Draw on a blank map of Europe, the track of the moving depression shown in Fig. 50.

217. (I.) Draw a diagram showing very clearly the changes in wind at Holyhead as the depression approached and passed away.

218. (H.) Describe from the charts of Fig. 50 how the rain in this depression spread across the country, and how fair weather followed the rain.

The approach of a depression after a spell of fine weather is sometimes heralded by the formation of high cirrus clouds (mares' tails). As the depression gets nearer, the sky may become overcast with cirro-stratus, through which the sun looks watery, and the moon pale. A halo may be formed. The clouds get thicker and lower, and the air becomes damp and muggy, as the depression gets nearer. Rain may fall, and continue to fall until the trough of the depression (line through the lowest barometer reading at right angles to the direction of travel, see Fig. 50) has passed. Then the weather may improve, becoming showery with bright sunny periods, for a while. Temperature frequently drops when the trough of a depression passes.

The weather at a particular observation station during the passage of a depression depends largely on the position of the station relative to the line of travel of the depression.

EXERCISES.

219. (E.) In *The Weather Map* by Sir Napier Shaw study especially the two examples of travelling depressions as illustrated by the charts in Plates VII and VIII.

220. (E.) Consider carefully the charts shown in Fig. 50. State how far the depression shown there had the distinctive features of a cyclone as given above.

Turn back to Fig. 46. You have there a type of pressure distribution very different from that of Fig. 44. In Fig. 46 it is the **highest** pressure which is in the centre and not the lowest. The pressure **decreases** as you travel outwards from the centre. Such a type of pressure distribution is called an **anticyclone**. Look at Fig. 51 which gives the isobars of Fig. 46 along with the winds, temperatures, and present and past weathers just as are

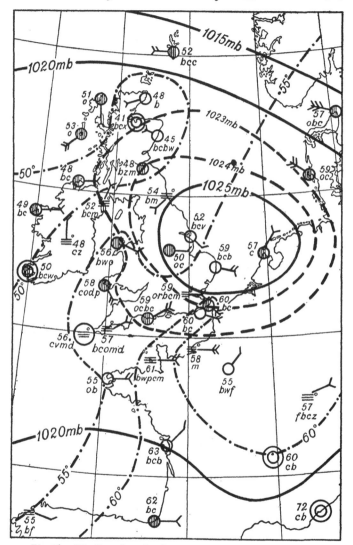

Fig. 51.   Weather chart for September 10th, 1917, 7 a.m. G.M.T.

given in Fig. 49. You will see from this example that an anticyclone is a region of calms or light winds, and fair weather. If you compare Figs. 48 and 51 you will at once notice how much more rain there is in the cyclone than in the anticyclone. Fig. 51 shows that some showers had fallen in the South of England, but everywhere else it shows fair weather. You will notice that mist is indicated at Scilly and Falmouth in Fig. 51. There is often a good deal of mist in an anticyclone.

A cyclone often moves very quickly, but an anticyclone nearly always moves quite slowly, sometimes remaining almost stationary for days.

The anticyclone is generally considered to be the fine weather type of pressure distribution. It is true that there is little rain in an anticyclone, but an anticyclone may be a cloudy one, so that there is no sunshine for days while the anticyclone persists. The cloudless anticyclone is one in which there may be a lot of fog, especially in autumn and winter.

The cyclone and the anticyclone are the two chief types of pressure distribution on a weather chart. There are other types which have their own characteristic weather, but the consideration of these must be left until you are able to carry on your study of the weather from a more advanced book than this one.

EXERCISE.

221. (H.) Obtain copies of the Daily Weather Report of the Meteorological Office. Use red and black inks diluted with water, and paint the isobar rings like contoured maps of hilly countries. Use deepening shades of red for over 1000 millibars, and deepening shades of grey for under 1000 millibars.

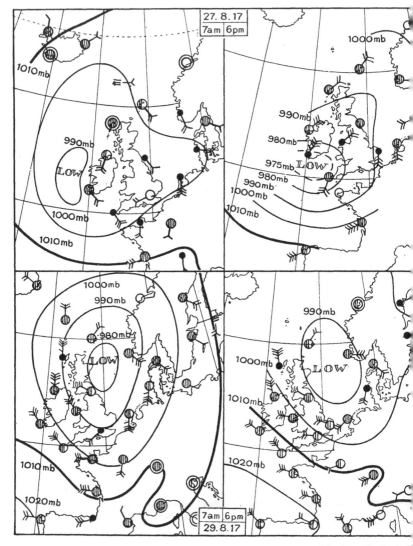

Fig. 50. Cyclonic d

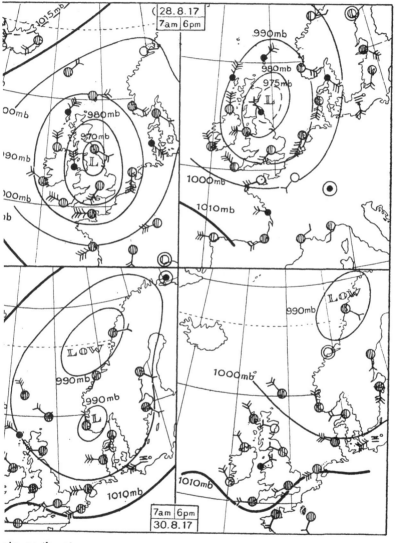

ving north-east.

CHAPTER X

ANTICIPATION OF WEATHER

IN the last chapter you have learnt something about the cyclone and the anticyclone, their movements and the weather they carry with them. You should now arrange to have the Daily Weather Report of the Meteorological Office, London, sent to you each day. If you have a wireless receiving installation you should pick up the Eiffel Tower weather reports daily. By the aid of the Daily Weather Report you will be able to watch the changes which are taking place, and to connect your weather with the weather over the British Isles. On the report indications of future weather are given, and you will be able to see how far such indications affect you. In other words you will be able to anticipate coming weather.

Perhaps the easiest case in which you can anticipate weather is when a cyclonic depression appears in the West and passes North-Eastwards or Eastwards over our islands.

EXERCISE.

222. (O.) Note the appearance of cirrus clouds at all times. Find out if they **always** indicate the approach of a depression.

The appearance of high cirrus clouds (see Fig. 52), especially the kind known as "mares' tails," is popularly believed to be a sign of rain. You can see the reason for this since "mares' tails" often indicate the approach of a depression, and a depression generally brings rain.

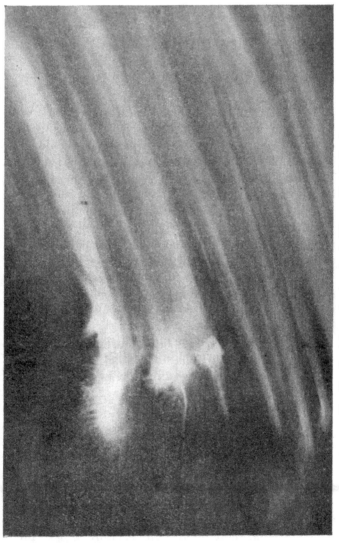

Fig. 52. True cirrus showing sheaf of fibres with tufted ends.

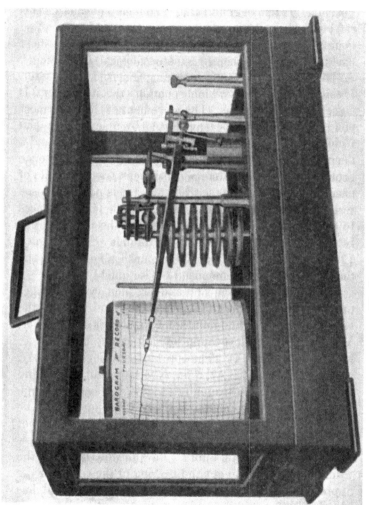

Fig. 53. Barograph.

When a depression approaches your station, your barometer will begin to fall. You may not notice this if you only read your barometer at fixed hours. You ought to have a barograph, which is one of the most useful and interesting of meteorological instruments. A barograph is a self-recording aneroid barometer. Inside the glass case which contains the instrument, is a set of aneroid boxes. The movement of the uppermost end of these boxes is transmitted by a lever to a pen which writes on a chart. This chart is wound round a drum which moves by clockwork, the drum going once round in a week. Thus a continuous trace is made of atmospheric pressure (see Fig. 55). It is possible to see when the barometer starts falling, or when it commences to rise. In Fig. 55 you can see that the barometer at Reading was falling all day on Monday 27th August, 1917, and that it went on falling until about 8 a.m. the next day when it commenced to rise quickly.

During the passage of a depression, wind changes are most likely to occur. When the wind shifts in the same direction as the hands of a watch it is said to **veer**. When it shifts in the opposite direction it is said to **back** (see Fig. 54). Thus a wind shifting from S.W. through W. to W.N.W. is a veering wind. A wind shifting from S.W. through S. to S.E. is a backing wind.

When your barometer begins to fall you should be on the lookout for a change of wind. If you cannot take frequent wind directions observe the motion of low clouds. From Buys Ballot's Law the direction of the wind enables you to get the bearing of the centre of the approaching depression. Suppose that your wind has been S.W. for a while, and that it begins to back towards S., and then to S.E., while the barometer is falling quickly, you would know that the approaching depression must have been to the West of you when your wind was S., and to the South-West of you when your wind was S.E.

If your wind backs from S.W. to S. then veers to S.W., W., and N.W., the centre of depression must have

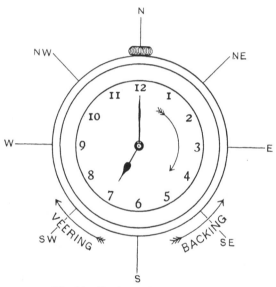

Fig. 54. Veering and backing winds.

been first to the West of you, then to the N.W., N., and N.E. of you. That is to say, the depression must have passed to the North of your station.

If the centre of a depression passed over your district you would observe a sudden shift of wind when the barometer stopped falling and commenced to rise. For example, the wind might be S.E. as the depression approached, and then suddenly shift to N.W. as the centre of the depression passed over you.

EXERCISES.

223. (O.) Whenever you are able to determine that a cyclonic depression is approaching your district take frequent wind observations and determine the path of the centre of the storm.

224. (O.) When a depression is approaching your station the barometer falls. Afterwards it commences to rise. Try to obtain wind directions and strength very frequently just about the time the barometer ceases to fall and commences to rise.

225. (O.) Take careful observations of wind and weather whenever a depression approaches and passes your station. Classify your notes according to the track of the centre of the storm. Determine very carefully the difference in the weather of a cyclonic depression whose centre passes to the North of you, and one whose centre passes to the South of you.

226. (E.) Study the Daily Weather Report of the Meteorological Office, London (issued about noon on the day of issue), and see if you can identify any depressions under whose influence you have been.

227. (O.) Take frequent observations of temperature, dry and wet bulb thermometers, on occasions of the passage of a depression.

228. (I.) The captain of a ship sailing from Ireland to Canada notices that his barometer is falling quickly. The wind which was South-Westerly veers to West. What course should the captain set in order to keep clear of the storm? Draw a diagram to illustrate your answer.

229. (I.) An Atlantic liner travelling due West passes through the centre of a depression which is moving North-East. Describe exactly what happens to a barometer on board the liner, and state what winds would be observed. Draw a diagram to help you.

230. (E.) Look up the nautical terms used in the following statement, and explain why it is true: In the Northern Hemisphere a rising barometer on the port tack is a valuable indication of improving weather, while a falling barometer on the starboard tack is a valuable warning of bad weather.

Turn back to Fig. 50 which illustrates the passage of a depression across the British Isles. Suppose that you had been at Reading on the days for which the charts in Fig. 50 are drawn, and suppose that you had been able to take frequent observations. At 7 a.m.[1] on the morning of the 27th August your wind would have been S.W. force 3, and there would have been an almost cloudless sky. As the morning went on you would have noticed a backing of the wind towards South, and a clouding over of the sky. You would have seen from your barograph (Fig. 55) that the barometer was falling continuously. Possibly you might have seen a solar halo. One was seen at Hartland Point at 10 a.m. that morning. You would have been quite sure that a depression was approaching from the West. During the afternoon you would have observed a further backing of the wind, and you would have noted that steady rain had commenced. At 6 p.m. your wind would have been E.S.E. and your weather " *or*." You might have been uncertain as to the track of the centre of depression. If you had been up by 6 a.m. the next morning you would have seen from your barograph that your barometer had ceased to fall (see Fig. 55). You might have observed a quick change in wind direction about this time. At 7 a.m. your wind would have been S.S.W. force 6, and your weather " *or*." Your rain-gauge would have told you that there had been 20 millimetres of rain since the previous morning. At 8 a.m. you would have seen that a distinct rise was taking place in the barometer. During the day your wind would have been S.S.W. or S.W. force 5 to 6, and your weather overcast with

[1] These times are all Greenwich Times, Summer Time is one hour later.

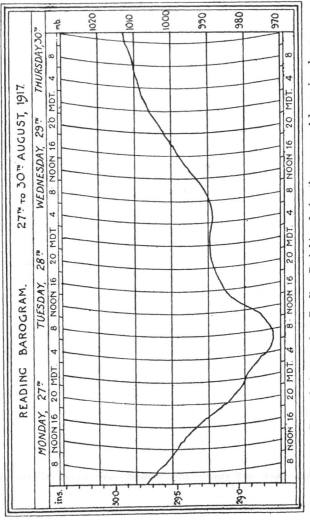

Fig. 55.   Trace of a barograph at Reading, Berkshire, during the passage of depression shown in charts of Fig. 50.

rain now and then. You might have seen patches of blue sky occasionally. On the morning of the 29th you would have been struck by the fact that the barometer had been steady from about 10 p.m. to 2 a.m., and had even fallen from 2 a.m. to 4 a.m. You might have suspected the approach of another depression had not your wind remained S.W. During the morning you would have seen your barometer rising steadily again, and as the day progressed you would have seen an improvement in the weather, and a veering of the wind, which by night had become Westerly.

This example will show you what a useful instrument a barograph is, and how very clearly it shows the changes which are taking place in atmospheric pressure.

The wind changes which take place on the approach of an anticyclone are not as noticeable as those which take place on the approach of a cyclonic depression. Often enough the winds in an anticyclone are so light that no definite direction can be assigned to them. The approach of an anticyclone would be indicated by a slow, steady rise in the barometer.

EXERCISES.

231. (E.) Trace a copy of the Reading barogram of Fig. 55. On this copy put wind and weather at the proper times from what has been said on pages 117 and 119.

232. (I.) What would be the wind changes for an anticyclone which travelled S.W. to N.E. and passed to the South of you ?

233. (E.) Whenever an anticyclone establishes itself over your district make careful notes of wind, weather, and temperature. Determine if the track of the centre makes any difference to the weather. Use your notes to anticipate coming weather when a new anticyclone is approaching your district.

In the last and present chapters you have dealt with the two chief forms of pressure distribution. There are many other and varied forms, and a knowledge of them leads an expert meteorologist to be able to forecast weather at all times from a series of successive weather charts. How this is done is explained in Sir Napier Shaw's book, *Forecasting Weather*. When you have completed your work from the present book, you will be able to continue your study of the weather from more advanced books, and you will certainly find that the subject grows more interesting the further you go into it. You will find some very wonderful things in the study of the weather, quite as wonderful as in any other branch of science.

APPENDIX I

REVISION AND SUPPLEMENTARY EXERCISES

234. (O.) Check the orientation of your observation.post by the Pole Star on a clear, starry night.

235. (E.) Obtain photographs of stunted trees in a bleak situation. How can you tell from them which is the prevailing wind for the district ?

236. (I.) Write out the letters of the alphabet. Mark those which are Beaufort letters and state what they stand for.

237. (H.) Read through Kingsley's *Westward Ho!*, and find the description of "The Wreck on the Shutter." Read also Dickens' *David Copperfield*, and find the description of the storm at Yarmouth. Give equivalent descriptions of these two storms, using only the Beaufort letters on p. 15, and the Beaufort wind arrows on p. 13.

238. (H.) Give the equivalent of Kingsley's poem, "Ode to the North-East Wind," using only Beaufort letters and wind arrows.

239. (E.) Make a collection of original cloud photographs.

240. (I.) Are these descriptions true to fact ?

    (a) "Rosy fingered dawn."

    (b) "The morn like lobster boiled from black to red began to turn."

    (c) "For a breeze of morning moves,
          And the planet of Love is on high,
        Beginning to faint in the light that she loves,
          On a bed of daffodil sky,
       To faint in the light of the sun she loves,
          To faint in his light and to die."

Write your own account of a sunrise.

241. (H.) Write a list of the colours in your paint-box. Name the colours or mixtures of colours you would use to represent (*a*) sky blue, (*b*) cloud grey, (*c*) sunset red, (*d*) daffodil sky, (*e*) aurora borealis. Paint samples of each of these.

242. (H.) On white paper paint in blue so as to make a picture of white clouds in a blue sky, and on blue paper use white paint or white chalk to make a cloud picture.

243. (E.) (i) Tennyson, in *Locksley Hall*, wrote :

"...a colour and a light
As I have seen the rosy red flushing in the northern night."

(ii) Longfellow, in *The Challenge of Thor*, wrote :

"The light thou beholdest
Stream through the heavens
In flashes of crimson,
Is but my red beard
Blown by the night wind,
Affrighting the nations."

(iii) Coleridge, in *The Ancient Mariner*, wrote :

"The upper air burst into life
A thousand fireflags sheen,
And to and fro and in and out
The wan stars danced between."

Do your observations agree with those of the poets ? Write your own description of anything of the sort which you yourself have witnessed.

244. (H.) Show by paints the three most marked colours that you see through a spectroscope or when a spectrum is thrown on the lantern screen. Paint the whole spectrum band from memory. When you next see a rainbow describe how its colours compare with those of the spectrum. Are they the same colours ? Are they in the same order ?

245. (E.) Devise some experimental means in the science laboratory to illustrate the formation of cloud, fog, dew and frost.

246. (H.) Write an essay on evaporation and condensation Describe in this essay the seven forms of condensation.

247. (L.) By experiment find out the amount of water-vapour in a litre of air. Repeat the experiment for several samples of air taken from your laboratory, and from out in the open on different days.

248. (V.) Discuss the monthly rainfall maps of the British Isles as given in the Monthly Weather Reports of the Meteorological Office.

249. (L.) Examine snow-flakes, hoar-frost and rime under a microscope. Make careful sketches of what you see.

250. (H.) Write your own description of any thunderstorm you have witnessed, then read what Ruskin says about Clouds in *Modern Painters*.

251. (O.) On the occasion of a thunderstorm, for each lightning flash take (i) the time the flash occurred, (ii) a compass bearing of the flash, and (iii) the distance of the flash (see Exercise 110). From your observations draw the track of the storm on a map of your county.

252. (L.) Prove experimentally that air has weight.

253. (V.) How is it that an aneroid barometer can be used as a height indicator on an aeroplane?

254. (H.) What is Buys Ballot's Law? How could you verify it apart from a weather chart?

255. (H.) Give a detailed description of a cyclone and an anti-cyclone. Emphasise especially the points of difference.

256. (E.) Describe how an aviator flying the Atlantic (i) from West to East, (ii) from East to West, should set his course with respect to a cyclonic depression which he may encounter in his flight.

257. (E.) Discuss the reports of previous day's weather as given in the morning newspapers.

258. (H.) From your observations check the official daily weather forecasts for your district.

APPENDIX II

The following, being Parts I and II of the *Syllabus of Weather Study for Elementary Schools* issued by the **Meteorological Office, South Kensington, London, S.W.**, is reproduced here by kind permission of Sir Napier Shaw, Sc.D., F.R.S., Director of the Meteorological Office. The work in this book covers these two parts of the syllabus.

METEOROLOGICAL OFFICE

SYLLABUS OF WEATHER STUDY FOR ELEMENTARY SCHOOLS

I. WATCHING THE WEATHER

1. How to tell the direction and force of the wind.

    Meaning of North, South, East and West, and the intermediate points of the compass.

    How to find out the North point, or one of the other cardinal points.

    How to set a weather-cock and to tell whether a weather-cock is properly set.

    How to give a number to the force of the wind by noting its effect on trees and buildings.

2. How to tell how hot or how cold it is.

    How to read a thermometer.

    How to set out a thermometer to tell the proper temperature in the shade.

    How to find out whether it makes any difference where you put the thermometer.

3. How to tell how hot it has been in the day and how cold it has
been in the night.

How to read a maximum thermometer.

How to put a thermometer so as to get the greatest heat in
the shade.

How to put a thermometer so as to get the greatest heat in
the sun.

How to find out whether there has been a frost on the
ground in the night.

4. How to find out whether the air is moist or dry (Sea-weed or
catgut hygroscope, wet bulb thermometer).

5. How to tell some of the different forms of clouds :

Woolpack clouds (cumulus), feathery clouds, mares' tails,
stripes of clouds, flocks of little clouds, mackerel sky
(cirrus), rain clouds (nimbus).

6. How to tell the way the clouds are moving.

7. How to tell how much rain has fallen.

How to set out a rain-gauge and use the measuring glass.

8 How to keep a proper record of the weather :

Hoar-frost, dew, fog, snow, hail, thunder and lightning,
rainbow, halo.

II. LOOKING OUT FOR CHANGES IN THE WEATHER

9. How to find out when the weather is changing.

How to read a simple mercury barometer, and how to read
a dial barometer.

How to tell whether the barometer is rising or falling.

10. How to know what the changes in the weather are going to be.

How to tell whether the wind has backed or veered.

How to know what weather to expect

    (1) when the wind is backing from the Westward to the Southward, and the barometer has begun to fall,

    (2) when the wind is veering from Southward to Westward and the barometer has begun to rise,

    (3) when the wind is backing from South to East and the barometer is falling,

    (4) when the wind veers from North to East and the barometer is falling.

11. How to find out whether other places have the same weather you have yourself.

12. How to find out the way in which weather travels.

13. How people can be told all over the country of the weather that is travelling towards them.

14. How people who are told that storms of wind or rain are expected to come towards them can watch for the first signs of their coming and tell whether the forecast is proving correct or not.

15. How people who know what the weather is all over the country can tell if the weather is likely to be settled or changeable.

APPENDIX III

LIST OF BOOKS

A. **Books giving instructions on the taking of observations and the management of instruments.**

The Observer's Handbook. Published annually. The Meteorological Office, South Kensington, London, S.W. 7. 3s.

Hints to Meteorological Observers. W. MARRIOTT. To be obtained from the Royal Meteorological Society, 70 Victoria Street, London, S.W. 1. 1s. 6d.

B. **Short popular books on Meteorology.**

Some Facts about the Weather. W. MARRIOTT. To be obtained from the Royal Meteorological Society. 6d.

Weather Science. R. G. K. LEMPFERT. (Jack.) 6d.

C. **Text-books on Meteorology.**

Lehrbuch der Meteorologie. Prof. J. HANN. (Tauchnitz, 1904.) 24s.

Handbuch der Klimatologie. Prof. J. HANN. (Engelhorn, 1908–11.)

Handbook of Climatology. Translated by Prof. R. de C. Ward from Hahn's *Klimatologie.* (Macmillan, 1903.) 12s. 6d.

A Popular Treatise on the Winds. W. FERREL. (Wiley, New York, 1889.) 16s.

Elementary Meteorology. R. H. SCOTT. (Kegan, Paul, Trench, Traubner & Co., 1893.) 5s.

Meteorology: The Elements of Weather and Climate. H. N. DICKSON. (Methuen, 1893.) 2s. 6d.

Modern Meteorology. F. WALDO. (American Book Co., New York, 1893.)

Elementary Meteorology. F. WALDO. (American Book Co., New York, 1896.)

Elementary Meteorology. W. M. DAVIS. (Ginn, 1894.) 10s.

Weather Influences. E. G. DEXTER. (Macmillan, 1904.) 8s. 6d.

Meteorology, Practical and Applied. Sir JOHN W. MOORE. (Rebman, 1910.) 10s. 6d. This book gives descriptions of old and new meteorological instruments. It describes the organisation of the meteorological services of the British Isles, the United States of America, and Canada.

Descriptive Meteorology. WILLIS L. MOORE. (Appleton, 1910.) 12s. 6d. An American book.

Barometer Manual. (Meteorological Office, London.) 1s.

The Seaman's Handbook of Meteorology. (Meteorological Office.) 3s. 6d.

The Weather Map. (Meteorological Office.) 4d.

The Weather of the British Coasts. (Meteorological Office.)

D. Books on Weather Forecasting.

Aids to the Study and Forecast of Weather. W. CLEMENT LEY. (Meteorological Office, 1880.) 1s.

Forecasting Weather. Sir NAPIER SHAW. (Constable, 1911.) 12s. 6d. This book contains a large collection of selected weather charts.

E. Books on Clouds, etc.

Clouds, their forms and combinations. ELIJAH WALTON. (Longmans, Green & Co., 1869.)

Cloudland. A Study of the Structure and Characteristics of Clouds. W. CLEMENT LEY. (Stanford, 1894.) 7s. 6d.

International Cloud Atlas. H. H. HILDEBRANDSSON and L. TEISSERENC DE BORT. (Gautier-Villars et Fils, Paris; second edition.) 10s.

Cloud Studies. A. W. CLAYDEN. (Murray, 1905.) 12s.

The Form of Clouds. Capt. C. J. P. CAVE, R.E. To be obtained from the Royal Meteorological Society. 2s. 1½d.

Cloud Forms. (Meteorological Office, London, 1918.)

Modern Painters. JOHN RUSKIN.

The Eruption of Krakatoa, and subsequent phenomena. Royal Society of London Report of the Krakatoa Committee. Edited by G. J. SYMONS. This publication contains frontispiece in colours of twilight and afterglow effects at Chelsea, London.

F. Books on Weather Lore.

Weather Lore. A Collection of Proverbs, Sayings and Rules concerning the Weather. R. INWARDS. (Elliot-Stock, 1898.) 7s. 6d.

A Medley of Weather Lore. M. E. S. WRIGHT. (Bournemouth, H. G. Common, 1913.)

A Handbook of Weather Folk-Lore. Rev. C. SWAINSON. (Blackwood, 1873.)

G. Glossary.

Meteorological Glossary. (Meteorological Office, London, 1918.) 1s. Contains short articles on the various terms in constant use in Meteorology. A most useful book.

H. Books and reports published periodically.

Daily Weather Report. (Meteorological Office, London.) 5s. per quarter, post free. Gives weather charts for 7 a.m. of day of issue, and for 7 a.m., 6 p.m. of previous day, barometer readings, temperature readings, rainfall etc. for many stations in N.W. Europe.

Weekly Weather Report. (Meteorological Office, London.) Annual subscription, 30s. Gives a weekly summary of weather over the British Isles.

Monthly Weather Report. (Meteorological Office, London.) Monthly, 6d. Gives a summary of the weather for the month of issue.

Symon's Meteorological Magazine. Edited by Dr H. R. MILL. Published monthly, 4d.

Quarterly Journal of the Royal Meteorological Society. Quarterly, 5s.

Journal of the Scottish Meteorological Society. Published annually, 12s. 6d.

INDEX

Air 74
Alto-cumulus 29, 30, 32, 40
Alto-stratus 29, 31, 32, 40
Aneroid barometer 79, 81, 88
Anticyclone 108, 110
Atmosphere 59, 69, 72, 75
Aurora 39

Backing winds 114, 115
Barograph 113, 114, 118
Barometer 77, 88
Beaufort, Admiral 12
   ,,   Scale of Wind Force 13, 82, 87
Beaufort weather notation 15, 17, 87
Ben Nevis 44
Books on Meteorology 127
Buys Ballot's Law 97, 105, 114

Catgut 66
Cave, Capt. C. J. P. 32
Celsius 61
Centigrade 60
Cirro-cumulus 26, 27, 28, 32, 40
Cirro-stratus 28, 29, 40, 108
Cirrus 19, 20, 21, 22, 24, 26, 28, 40, 111, 112
Clouds 18–34, 40, 41
Cobalt chloride 67
Colours of the sky 35
Compass 6, 8
Condensation 41, 42, 47
Coronas 36
Cumulo-nimbus 23, 41
Cumulus 20, 21, 23, 25, 26, 40
Cyclone 105, 107, 110

Daily Weather Report 86, 89, 111
Depression 105, 111
Dew 45
Dry bulb thermometer 68, 69

Evaporation 41, 49, 68, 69

Fahrenheit 60
Fir cones 66
Fluid 73
Fog 16, 41–44, 110
Fracto-cumulus 41
Frost 45

Glycerine barometer 79

Hail 48
Hair hygroscope 67
Halo 36, 37, 108, 117
Hoar Frost 45
Horizontal rainbow 39
Howard, Luke 22
Human hair 66
Humidity 65, 68

Isobar 93, 107
Isotherm 98

"Jacky and Jenny" 66

Kew barometer 79, 80, 83, 89

Lightning 57, 58
Lunar corona 37

Mackerel sky 26, 27
"Mares' tails" 22, 108, 111

Maximum thermometer 70
Mean sea-level (M.S.L.) 84, 89
Meteorological Office 26, 86, 87, 89, 111
Metric system 81
Millibar 82, 93
Minimum thermometer 71
Mist 16, 41–44

Nimbus 22, 25, 26, 33, 40
Nitrogen 72

Observation book 17, 84, 85
Observer's Handbook 39, 62
Orientation of observation post 7, 10
Oxygen 72

Pressure of atmosphere 72, 76, 79–81
Pressure units 81

Quarterly Journal of the Royal Meteorological Society 32

Rain 47
Rainbow 36
Rainfall 48, 51–54
Rain-gauge 49–51
Revision and Supplementary Exercises 121

Sea-level pressure of atmosphere 81
Seaweed 65

Shaw, Sir Napier 120, 124
Ship's screen 62, 64
Snow 48
Snowfall, measurement of 55
Soft hail 48
Solar corona 37
  ,,   halo 38
Stevenson screen 62, 63
Strato-cumulus 29, 32, 33, 34, 40
Stratus 21, 24, 26, 33, 40
Submarine 74, 75
Syllabus of Weather Study 124

"Table-cloth," Table Mountain 44
Temperature 60, 62, 68
Thermometer 60, 61, 68
  ,,   screen 68
Thunderstorms 56, 58
Torricelli 77

Veering winds 114, 115
Vernier 83, 84

Water 73
Water barometer 79
  ,,   vapour 41, 42, 43, 65
Weather charts 86
  ,,   diary 1, 2
Weather Map, The 88
Weather signs 2, 3
Wet bulb thermometer 68, 69
Wind direction 5, 10
  ,,   strength of 12
  ,,   vane 11

Printed in the United States
By Bookmasters